Properties of Matter

TEACHING CHEMISTRY IN A DIVERSIFIED CLASSROOM

BOOK 2

By Sammie Jacobs

Copyright

All rights reserved under International and Pan-American Copyright Conventions. By payment of the required fees, you have been granted the non-exclusive, non-transferable right to use the materials provided for the preparation of lessons and during the direct instruction of students.

PROPERTIES OF MATTER. Copyright 2020 ©

Student worksheets and assessments may be duplicated for classroom use, the number not to exceed the number of students in each class. Notice of copyright must appear on all copies as provided on the page.

Activity kits duplicated and assembled as directed should display the copyright markings as prescribed in the directions and on the materials.

No other part of this text may be reproduced, up-loaded, displayed, shared, transmitted, down-loaded, decompiled, reverse engineered, or stored in or introduced into any information storage and retrieval system, in any form or by any means, whether electronic or mechanical, now known or hereinafter invented, without the express written permission of Lavish Publishing, LLC.

First Edition
2020 Lavish Publishing, LLC
Teaching Chemistry in a Diversified Classroom book 2
All Rights Reserved
Published in the United States by Lavish Publishing, LLC, Midland, Texas
Select Illustrations by Sanghamitra Dasgupta
Cover Design by: Victor R. Sosa
Cover Images: CanStock
Paperback edition
ISBN: 978-1-64900-001-9
www.LavishPublishing.com

FOREWORD

Teachers, parents, and educators of all kinds – welcome to my classroom!

Here within these pages, and in fact this entire series, are lessons and activities that I have created and used with my own students over the years presented in an easy-to-use format. Through trial and error, hardship and success, I have learned how to present the difficult world of introductory chemistry to students in scalable terms, ways that lend themselves to a wide variety of learners.

So, if you teach in a small class, large class, average class, have ESL or SPED, or even if you are homeschooling, you will find the flexibility of my lesson plans tailor made for you. Each unit includes a complete calendar plan, lessons for each classroom day for forty-five minutes of instruction each, and even formative and summative assessments to check for learning. Simply use the lessons, scale them as need be, and augment whenever you like.

Be sure to join my Teaching Chemistry PLC on Facebook for even more ideas and sharing, and of course, I hope for the best with you and your students!

Sammie Jacobs

TABLE OF CONTENTS

Properties of Matter Unit Tools ..... Pages 1 – 5
Calendar Key and Usage
Properties of Matter Guide Calendar
Properties of Matter Blank Calendar
Properties of Matter Vocabulary

Lesson Plans .... Pages 6 – 77
Chemical and Physical Change
Chemical and Physical Properties
Describing Matter
Precision and Accuracy
Soda Lab (Density)
Reading Day
Matter Stations Lab Activity
Element or Compound
Pure Substance or Mixture
Heterogeneous and Homogeneous
Matter Flow Chart
Properties of Matter Review

Assessments .... Pages 78 – 105
Property or Change Quiz
Chemical or Physical Quiz
Intensive or Extensive Quiz
Quantitative or Qualitative Quiz

Precision and Accuracy Quiz
Density Quiz
Element or Compound Quiz
Pure Substance or Mixture Quiz
Homogeneous or Heterogeneous Quiz
Matter Flow Chart Quiz
Properties of Matter Vocabulary Quiz
Properties of Matter Unit Exam
Answer Keys

Unit Calendar

The **Unit Calendar** is an important organizational tool.

It features a list of vocabulary terms (top section) for the unit, the plan of days (center section), and the student learning goals (bottom section).

A **completed teacher copy** of the calendar with the unit overview is provided. On this version, days are not written as dates, but as lesson days. However, on my copy that I use in my classroom, these have been laid to the school or district calendar and those day numbers are replaced with actual meeting calendar dates. There are five in a row, which is a work week, and days that we do not meet or have instruction time are crossed off.

You will notice that the unit calendars are always a whole number of weeks long and every day has a lesson planned, but often you will have holidays or non-class days. When you have less days, combine lessons where appropriate, or omit lessons that are not included on the student goals and exam if necessary. We also only meet for forty-five minutes per day, so the lessons are designed to fit within that time frame. If you have longer class periods, you may decide to augment the days with additional activities, such as vocabulary games.

The **blank student copy** can be used to customize the calendar to match your available days. It can also be copied and distributed to students, along with the vocabulary lists with definitions.

Calendar ideas:
1. Have students fill in the calendar daily with daily topic as they begin class as a way of making them aware of what the day's lesson will entail.
2. Have students fill out the calendar at the start of a new unit so they have the plan ahead of time and can be proactive with their learning.
3. Give students a completed copy with the appropriate dates and topics at the start of a new unit to save time.
4. Have students highlight each day and the corresponding terms on the vocab list with a different color (color coordinating them) for reviewing later.
5. Have students use the calendar to record assignments for each day and highlight them as they are completed and submitted.
6. Have the students write a question for each day to reflect upon later.

Unit	Properties and Changes of Matter	Month

Vocabulary

accepted value	heterogeneous	precision
accuracy	homogeneous	pure substance
chemical change indicators	intensive	qualitative
chemical property	malleability	quantitative
compound	mass	reactivity
Density	percent error	solubility
Ductility	phase change	suspended
element	physical property	toxicity
extensive	precipitate	volume

Daily Agenda

M	T	W	Th	F
1	2	3	4	5
Chemical and Physical Change	Chemical and Physical Properties	Describing Matter	Precision and Accuracy	Soda Lab

M	T	W	Th	F
6	7	8	9	10
Reading Day	Matter Stations Day 1	Matter Stations Day 2	Matter Stations Day 3	Element or Compound

M	T	W	Th	F
11	12	13	14	15
Pure Substance Or Mixture	Homogeneous Or Heterogeneous	Matter Flow Chart	Flex and Review day	Properties of Matter Exam

Learning Goals

Target Concept	none	weak	solid
Differentiate between physical and chemical changes and properties.			
Differentiate between intensive and extensive properties.			
Differentiate between qualitative and quantitative properties			
Classify measurements as precise and / or accurate.			
Classify matter as pure substances or mixtures			
Classify pure substances as elements or compounds			
Classify mixtures as heterogenous or homogeneous			

© Lavish Publishing, LLC

Unit **Properties and Changes of Matter**   Month _____

Vocabulary

accepted value	heterogeneous	precision
accuracy	homogeneous	pure substance
chemical change indicators	intensive	qualitative
chemical property	malleability	quantitative
compound	mass	reactivity
density	percent error	solubility
ductility	phase change	suspended
element	physical property	toxicity
extensive	precipitate	volume

Daily Agenda

M	T	W	Th	F

M	T	W	Th	F

M	T	W	Th	F

Learning Goals

Target Concept	none	weak	solid
Differentiate between physical and chemical changes and properties.			
Differentiate between intensive and extensive properties.			
Differentiate between qualitative and quantitative properties			
Classify measurements as precise and / or accurate.			
Classify matter as pure substances or mixtures			
Classify pure substances as elements or compounds			
Classify mixtures as heterogenous or homogeneous			

© Lavish Publishing, LLC

Properties of Matter Vocabulary

word	definition
accepted value	The book value or one given to you by you teacher is called the ___; it is what we should be getting in our measurements.
accuracy	Measurements close to what they should be is called _____.
chemical change indicators	There are four ____. They are: formation of a gas (bubbles or odor); color change, temperature change, or formation of a precipitate.
chemical property	A ___ cannot be observed without changing the composition of the substance; examples are flammability and reactivity.
compound	A _____ is 2 or more different elements chemically combined.
Density	_____ is an intensive property of a substance that is measured in g per mL. It is a factor that determines whether an object will sink or float when placed in a liquid.
Ductility	____ is the ability to draw metal into wire.
Element	An ___ is matter in its simplest form.
extensive	An _____ property changes with the size of the sample, like volume.
heterogeneous	____ mixtures have varied amounts when sampled and are physically combined.
homogeneous	____ mixtures have unvaried amounts when sampled and are separate physically.
intensive	An ____ property stays the same regardless of sample size - like density.
malleability	_____ is the ability of metal to be shaped or flattened.
Mass	____ is the amount of matter present - constant regardless of location in the universe.
Matter	___ has mass and volume.
percent error	___ is calculated to determine how close or far our results are from where they should be based on the accepted value.
phase change	A change from one state (solid or liquid or gas) to another without a change in chemical composition is called a ____.
physical property	A ____ property can be observed without changing the composition of the sample.

precipitate	A ___ is a solid formed from combining two liquids - indicates there was a chemical reaction.
precision	Taking experimental values that are close to each other is ___.
pure substance	Examples of ___ would be a cup of sugar, a flask of HCL, or a bottle of O_2, which means only a single substance is present.
qualitative	___ properties or values can NOT be given number form - such as color.
quantitative	___ properties or values ARE given number form - such as mass.
reactivity	The willingness of an element to form bonds is called ___.
solubility	___ is a physical property referring to the ability of a given substance, the solute, to dissolve in a solvent. It is measured in terms of the maximum amount of solute dissolved in a solvent at equilibrium.
suspended	When an object is placed in a liquid, but it does not float on top or sink to the bottom, it might appear that is it just hanging in the middle somewhere - this object would be said to be ___ in the liquid.
toxicity	___ is the chemical property of a compound that is poisonous or harmful if ingested (consumed).
volume	___ is the amount of space matter occupies.

© Lavish Publishing, LLC

Key to Lesson Plans

Topic: coordinated to the unit calendar - what we are learning about today?
Day: coordinated to the calendar - sequence of presentation out of total available
Unit: coordinated to the unit calendar - unit the lesson falls under

Learning Target:
A postable **learning objective** that gives an outline of what the day's lesson will cover and **artifact** the students will produce for evaluation. These can be used as is or modified to suit your student output or to focus on a different aspect of the lesson, as these often only cover one part of a layered lesson and group of activities.

Student Goals:
At the bottom of the calendar, the students have a list of **goals** they want to meet by the end of the unit, which are covered on the **unit exam**. This section coordinates which goals are being addressed during this lesson. This section also lists the **vocabulary terms** from the calendar that will be defined or needed during the lesson so they can be connected, pre-taught, or directed to study afterwards.

Agenda:
A postable list of what activities this day will include. They are divided into two main types of instruction:

Lecture **(L)**, where the teacher is providing direct instruction and the students are actively listening, taking notes and providing feedback at given intervals.

Activities **(A)**, where the student is producing work of some kind and the teacher is observing and providing support when needed.

Student Materials:
A list of materials that each student, pair, or group, will need to complete the lesson. They will need to be prepared ahead of time and be ready to **pick up** as students enter the room or to be **distributed** at the appropriate time during the lesson.

Generally, if it is a **one-to-one item**, I have them ready and students pick them up as they come in to save time. **Paired and Group materials** can be handed out or placed in a location to have a student go and get while the teacher is preparing some other part or completing a task during transition.

Props:
These are items you generally only need one of and can be hidden to pull out at the appropriate time or placed on a table that the students can visit before and after the lesson.

They are support items that deepen the understanding of the lecture and either generate questions or provide answers.

Having props is vital for a wide range of learners, so do not feel limited to what I have and use – explore and add anything that you want to aid in this process, as all students benefit. Save your props as you build them, as they often are used in future lessons to bring concepts forward and tie ideas and understanding together.

I have a set of props that stay on my front table throughout the year, as I pick them up and refer to them often: a plain bottle of water that is labeled H_2O, a bottle that is half cooking oil and half water, a bottle that is water with about a gram of dirt in it, and a bottle that is water with about a gram of corn starch in it. Other props are added and removed as needed.

Actions and Rationale:
These are the key points of the lesson and the reasoning behind what is being said or done. All activities will have a separate directions page if needed, as well as a reproduceable student copy and directions for building all props and student materials.

During each unit, we use a variety of learning and presentation styles that include having the students listen, speak, read and write. There are also a variety of study and practice skills woven into the lessons that students can learn to use in and outside of their Chemistry class.

Topic: Chemical and Physical Change
Day: 1
Unit: Properties of Matter

Learning Target:
I can review how to tell the difference between Chemical and Physical change and practice with examples.

Student Goals:
Students typically begin learning about Chemical and Physical Change early, so today is a chance to review and to narrow in on one detail – CHANGE.

Agenda:
A – Warmup: States of Matter Flow Chart
L – Chemical Change Indicators
A – Chemical and Physical magic boards

Student Materials:
Journals for notes
Calendar and Vocab copies for Gluing
Chem / Phys magic boards
Dry erase markers and erasers

Props:
Partially filled states of matter flowchart for warmup
A plain white piece of paper you can fold or tear if you want

Actions and Rationale:
Warmup – calendar and flowchart. My students copy the calendar from the doc-cam and glue them in their journal. I can make any related announcements before we start or at the end of class. For the flowchart, I display a partially filled in copy (solid, liquid, and gas with the arrows). They copy it in their journal and speculate with their partner what the rest is going to be. They fill out the rest as you go over and review. If you have ESL students, be sure to take your time – this is a review activity, but like new to those who learned this in another language.

Lecture – Once you have the chart filled in, tell the students that this is a review of Chemical and physical **CHANGE**, and today we are going to really focus in on what that word means. Then ask, "So how do you know when a CHANGE has taken place?" Give a few a chance to speculate, then say, "So how about in advertising, when they want to sell you something. Like a carpet cleaner or a diet program? What do they show you to convince you?" Give another one or two – someone may get it now, as we are going for BEFORE and AFTER. That is the key to change – there must be a before version and after version that are different.

Once you have established what change means, go back to your warmup chart and give it a few examples of pointing and explaining – solid to liquid, the before and the after. The gas to liquid, the before and the after. Be dramatic.

Then ask – "Do you think this is true for chemical changes?" You can let them talk big group or have them partner talk on this one. After about a minute, call them back to order and move to a new location in their journals to take notes over chemical change. I like to put this on the board near our **physical chart**, which is what I label the states of matter flowchart, and I tell them, "Everything on this chart is physical. Always. If it is a word on here, it's a physical thing."

Then I move over and make a new chart for the chemical, which I label Chemical Change Indicators. I use two colors – one for the indicator and one for what it looks/feels/sounds like when we **experience** it. We build it together, giving them a chance to remember what they have learned in the past, and then adding the experience part. When we are done, I say again, "This is the chemical chart. If it's on this chart, it indicates that it COULD be chemical, but we can't say for sure because chemical stuff is tricky."

Chemical Change Indicators

Gas produced – bubbles, odor, smoke, sizzle

Unusual Color change – not expected (makes the wrong color)

Energy Change – energy change (hot or cold feel or light emission)

Precipitate formed – two liquids that form a solid

Then we talk about bubbles. Does boiling water bubble? Sure it does. Does that mean it's chemical? No, the bubbles are there because we are adding heat to make it boil – and boil (vaporize) is on the physical change chart – that means it **has to be** physical. What if we pour two chemicals together and they bubble, but we aren't adding any heat?

I give them a chance to reflect and talk about that, then we decide – if it comes from mixing two substances (or chemicals) and we don't do anything else to it, and it shows up on our list, it's a good indicator that what happened was chemical. That's the before and after of CHEMICAL CHANGE – the BEFORE is the two different / separate chemicals, and the AFTER is the new stuff we are getting, evidenced by the chemical change indicator we are experiencing.

Now, many students have been taught the difference between chemical and physical is "Can I get it back?" or "Did you get something new?" so I like to address that, as one of those ways is better than the other. We even write DID YOU GET SOMETHING NEW? on the edge of our notes and mark it with a box or a cloud around it. Then, we take a few minutes to talk about what something new really means – that the chemical structure of the material has been altered.

Go back to the states chart – if it's water as a solid (H_2O), what is it when it becomes a liquid? The students (some if not all) should realize that it is still water. If they seem lost, do a few more on the chart, but keep using water, until they understand that it may LOOK different, but the composition is the same everywhere on the physical board. **Note:** the physical board is not all the ways to get physical change – there are many more, incase you want to point that out as you transition into the next step.

When they have the physical on the chart, I move on. I hold up an actual piece of paper and ask, "What about this? If I fold it, is it still paper?" Fold the paper and let them decide. "What if I tear it up?" You can actually tear it if you want. "Can I put it back as a whole sheet?" They are going to say no. "But, is it still paper?" Yes, it is still paper. Then I explain that this is why I don't like the question "Can I get it back" and that "Did you get something new" is a better question – talking about the new being at a molecular level. Finally, you can ask, "So, what if I burn it?" That's chemical change – ashes and smoke are something new at the molecular level and we get heat and light (chemical change indicators from our list). Lastly, you should point that out – the before is the paper, and the ashes and gasses produced are the after.

Activity – magic boards. This is our first time to use these, but they make learning so much less stressful for the students. You should invest in a box of smooth finished page protectors (the 'non glare' aren't good for this) and a ream of multi-color card stock. I have a permanent set for each unit, but that gets expensive. You can trade them out as you go and save the pages for next year while reusing the plastic part. You will also need a supply of dry erase markers and an eraser (I use wash rags that they share, or you can cut up a towel).

To set up for the activity, you run your worksheet on the colored card stock and insert them into the page protectors. These and the dry erase markers and erasers can go on the pick-up table as they come in, or someone from each group can pick them up when the time comes.

Running the activity gives you lots of options:

~ You can work through as a group, hitting and checking together

~ You can have them work with their partner and talk about what they are choosing and why

~ You can have them work separately and check their work against their partner's, reconciling who is right when they have different answers

~ You can have the answers to the page posted on your 'answer board' and they can work a few (or all) and then go check their work – if they are all right it's high fives, and if they are missing some, they need to figure out why (resources: notes, partner, group, or you)

Now, let's talk about why this works. It works because when you are putting answers down on a piece of paper you are going to turn in, you have to be RIGHT right now. If you put it on an erasable board, that means you are allowed to be wrong because you are learning and you are going to erase it when you are done. Plus, there's no reason to cheat. That's less stress. And maybe even a little fun. I like to call this **penalty free practice** – there are no points to be earned and no grade going in the book. It's all about the learning, and this is where the students get to rate themselves on that chart at the bottom of their calendar – if they are getting all right, they are solid, missing a few means weak, and missing lots means none. In the end, they are self-monitoring and learning to be responsible for their own education.

Chemical or Physical Change

Imagine the BEFORE and the AFTER – decide if each is a chemical or physical change.

1.		A bicycle rusting after it is left in the rain
2.		A pot of water boiling
3.		A sugar cube dissolving in a glass of water
4.		An iron nail rusting
5.		Baking a cake
6.		Bending wire
7.		Boiling
8.		Breaking a glass bottle
9.		Burning of wood
10.		Candle burning
11.		Candle melting
12.		Chewing of food
13.		Chopping wood
14.		Combining two colors of play-doh
15.		Combustion
16.		Compounds are broken down into elements
17.		Condensing
18.		Corn being ground
19.		Corrosion
20.		Cutting an apple
21.		Cutting grass
22.		Decomposition of a fallen tree
23.		Decorating a cake
24.		Deposition
25.		Digestion of food

© Lavish Publishing, LLC

Chemical or Physical Change

Imagine the BEFORE and the AFTER – decide if each is a chemical or physical change.

1.		Dying Easter Eggs
2.		Dynamite going off
3.		Elements combine to form compounds
4.		Erosion
5.		Evaporating
6.		Fireworks exploding
7.		Freezing
8.		Frying an egg
9.		Glow Stick
10.		Ice forming when water is placed in a freezer
11.		Knitting a sweater
12.		Melting
13.		Molding clay
14.		Oxidation
15.		Photosynthesis
16.		Respiration
17.		Ripping Paper
18.		Rusting
19.		Smashing
20.		Souring of milk
21.		Sublimation
22.		Sugar dissolving in tea
23.		Tarnishing
24.		Watercolor paint drying on paper
25.		Weathering

© Lavish Publishing, LLC

Topic: Chemical and Physical Properties
Day: 2
Unit: Properties of Matter

Learning Target:
I can learn how to tell the difference between Chemical and Physical Properties and practice with examples.

Student Goals:
Differentiate between physical and chemical properties

Agenda:
A – Warmup: Chemical or Physical Quiz
L – Properties versus Changes
A – Property card sort

Student Materials:
Journals for notes
Chemical or physical quiz slips
Card Sort Cards

Props:
Notes from Chemical and Physical Change
A discussion item (can be anything with obvious chemical and physical properties)

Actions and Rationale:
Warmup – chemical or physical quiz. Have the students take the quiz at the start of class. I like to take them up for practice points, 1, 2 or 3. Then we go over it – the first question is physical (it's a property, too), the second chemical (that's a change), and they get to define the difference, so you get a chance to remind them of our essential question, "Did we get something new at the molecular level?" Keep that in mind as we move into properties.

Lecture – properties versus changes. Today we are looking at Chemical and Physical, which is related to yesterday, but deeper than that, we need to know the difference between a Property and a Change. So, review how we know something is a change. Students can confer with their partner, and then someone (volunteer or random call) can define – Changes have a before and an after.

Ask, "So, what does property mean?" Give them a minute to talk about how to define property. They almost always come up with 'something you own' which is exactly what we will be talking about. Properties BELONG to the sample, so there is no before and after – they just are what they are, like a point in time. Is it or can it… whatever the property is.

Pull out your discussion item and place it in a location that everyone can see it. Have the students start calling out properties of the item – you can make a list on the board. You may find they mostly give you physical descriptors, but if you prompt them, they may come up with some chemical ones.

Once you have a little list of properties, focus in on how you test those properties. For example, if it's color – you look at it and observe. Was the sample hurt? No, the sample is fine. Mass – grab a scale and measure it. Hurt? No, it's fine. Flammability? If it IS flammable, what happens to the sample? Yup. It gets destroyed because it burned up. How do we test toxicity? Eek! Who wants to test that one? No one, because it would make the person sick or worse if it's true! (Yeah, I make it a little funny, but the kids love it.)

The point is, observing physical properties – the sample is fine. Testing chemical properties, the positive reaction destroys the sample. So that is how we can differentiate between them.

Now we have to point out the big picture. We have chemical and physical, and we have property and change – which means we actually have FOUR things – chemical property, chemical change, physical property and physical change. That means we have two ways we can figure out what things are… looking at the first parts (chem or phys) or the last part (prop or change).

Activity – practice sort for properties. You can copy them onto your colored card stock, cut them up and put them in baggies ahead of time. If you laminate them before you cut, they will last longer as well. And again, you have some options how you work them… whole class, small groups, or partners. After they have their boards set, walk around and give them a quick look for accuracy, then discuss what they chose.

~ Which side has more? **properties**

~ Which could we put on our states of mater board? **Melting and boiling points**

~ Which ones were hard to decide? **Point out any you saw that were on the wrong side to talk about**

Property Card Sort – copy, laminate and cut (colored cardstock or paper works)

Physical Properties   Odor

Chemical Properties   Taste

Flammability   Transparency

Hardness   Ability to Rust

Texture   Ability to Tarnish

Density   Number of Protons

Melting Point   Boiling Point

Number of Electrons   Color

Alters the material to test

Doesn't alter the material to test

Topic: Describing Matter
Day: 3
Unit: Properties of Matter

Learning Target:
I can learn and practice using some key types of descriptors of matter.

Student Goals:
Differentiate between qualitative and quantitative properties
Identify extensive and intensive properties

Agenda:
A – Warmup: Property and Change Quiz
L – Flip sides 4 corner notes
A – Partner Practice

Student Materials:
Journals for notes
Slips of paper for the quiz
Property magic boards
Dry-erase markers and erasers

Props:
A set of drink containers of some kind (I use bottles of water) in two sizes, a large and a small, that are otherwise identical. You will also need a picture of something in two sizes for them to look at and list properties of – I use a picture of lemonade in a pitcher and a glass of lemonade beside it.

Actions and Rationale:
Warmup – property and change quiz. This should take five to eight minutes for them to take and go over. I usually take them up for data, and then I present the answers as a quick reteach of the difference between a change and a property. Then I transition into the lesson, which is focusing on different types of properties that we use to describe matter.

Lecture – flip sides. This is the first time that we are using "4-corner notes", so I give them a quick explanation of how to divide their page so it looks like an old-timey kitchen window with four panes – which is to draw a large cross on their page, creating four boxes. We will use this style many times through the year.

The students put the topics in the top of each box and write a short definition under it, then leave the bottom half of each box open for examples, pictures, etc. You can give them the box order and have them copy the definitions from the vocab list if you want before you begin explaining, as they listen and participate better when they aren't writing. It also gives you a chance to set up your board – I usually make two T-charts side by side. One is labeled Quantitative and Qualitative and the other is labeled Intensive and Extensive. Place the smaller container of your props out to start your discussion.

Once everyone is set up, take a moment to define "flip sides of a coin." I like to be dramatic and usually hold an actual quarter up as I ask, "Have any of you ever flipped a coin to decide something?" I follow with, "Can both sides win?" This will need clarification, because some will usually be thinking that either side COULD win means the same thing, but it doesn't. Clarify – when you flip a coin, only one side WILL win – it will land either on heads or tails, but it can't be both at the same time.

That is how it is with today's properties – they are either Quantitative OR Qualitative. They are either Intensive OR Extensive. Each of these sets are exclusive against each other, like flipping two coins at the same time. When we choose a property, it can only appear once on each 'T-Chart' that we have created. Be dramatic and point at your bottle (mine is water) and say, "I can say that my bottle is 'plastic' in composition. Plastic can only be on one side on this chart" – indicate the Q&Q chart, "And it can only be on one side of this chart" – indicate the I&E chart. "It gets to be on both charts, but only on one side or the other of each. So, the Q&Q are on opposites side of the coin, one or the other, but not both will win."

Most of the kids will get this, so I don't spend more time here – we just go into examples and revisit after the lecture. By then, those who didn't get it will have time to process and see examples, which will help in their understanding.

Continue, either by giving an example or two before you move to student suggestions or jump straight in and have them suggest things to go on the Q&Q chart. For example, they use a 'size' descriptor – clarify that (large and small is not a size, but actual mass and volume are). Tell them to be specific. Remind them that a property is something you own or that belongs to you. The bottle might be 20 oz – 20 oz is a property of that bottle – so is it Qualitative or Quantitative? Obviously, it is quantitative – we assign a counted (measured) number to it – and we write it in that column.

It tastes "Sweet" is another possible property – qualitative – as that is not a number (counted or measured). Again, remind them that the word taste is NOT a property, but that how it tastes is. If they still don't get that part, I say, "Is the word mass specific?" No. It's a generalization that describes anything because all things have mass. "What if I say this marker weighs 10 grams?" and I hold up my marker. That is a specific property of a sample or item. That's what we want – we can generalize and know where certain types of values are going to fall, but in the end, we want specific properties of a given item or sample.

From there, continue on and cover extensive and intensive in the same way. Pull out your second prop – the identical one that is of a larger size. Tell them to look at the two and compare. The items that go on the I&E chart are the ones that were either changed or unchanged by having a larger sample.

Take some of the properties from the Q&Q chart – such as the volume. Which side will it go on for the I&E chart? Again, remind them that volume is generic, and 20 to 48 oz is specific – and they are Extensive. Revisit the 'taste' or 'color'. Again, be specific, and they are Intensive. Depending on the quickness of your students this can be very deep and may go quickly or take a while.

Activity – partner practice. Once you feel a large portion of the students understand the concepts, give them a fresh set of items to list properties of – I use a picture of a pitcher of lemonade next to a glass of lemonade. This can be projected on the whiteboard, hung on the whiteboard, or smaller copies to distribute to their desks.

For the rest of class, they work as partners to build a set of T-charts for the lemonade. Then, after about five minutes, we compare with our table, which is the full set of four (two partner pairs) and they get to talk about what they chose – and decide if they were correct. Any discrepancies need to be resolved before they clean their boards and turn them in. Or, if you have more time than that, you can even pull out responses into the full class group and discuss them.

Topic: Precision and Accuracy
Day: 4
Unit: Properties of Matter

Learning Target:
I can verbally and visually express the difference between precision and accuracy with a group activity.

Student Goals:
Classify measurements as precise and / or accurate.

Agenda:
A – Warmup: A&P Paragraph
L – Accuracy and Precision Story
A – Team Target Practice
A – Notes Writing

Student Materials:
Journals for notes
Slips of paper for the warmup
Target sets for activity

Props:
A story about being accurate and precise

Actions and Rationale:

Warmup – A & P paragraph. Have the students start off with putting their thoughts into words with the prompt, such as: "What is the difference between being accurate and being precise?" I tell them how much, like "give me three to five sentences." They write while I take attendance, then I collect them for points, and we talk about their responses to kick off the lecture. Often, they will say that they are the same, or try to use some other differentiation. I clarify with the definitions – accuracy is hitting the accepted value (book value or what the measurements should be, called the target) and precision (taking measurements that are close to each other and not all over the board).

Lecture – accuracy and precision story. I like to share stories about my sons – they help me express ideas and share a bit about myself with my students along the year. For this one, I tell them about a time my boys and I were at a camp one summer. At the camp they had a basketball goal, but there was no wall behind the goal – there was an empty half acre field. Every time we threw the ball and missed, we had to chase it into the field. Then I give them numbers and draw on the board – a basketball goal with arrows. My oldest son threw the best – say 100 times with 90 going into the goal. Was he accurate? Was he precise? Give those some discussion, then nail it down – he was both – he hit the goal and his shots were close to each other.

My second son, who was only twelve at the time, he wasn't so good. Bless his heart, he threw the ball 100 times as well, but being vertically challenged, he hit the pole under the basket 85 times. Was he accurate? Was he precise? Again, let the discussion flow – he was not accurate, obviously, as the goal is to make the basket. He is however precise – the is hitting close together. I go further and share that this is what happens when you work with a broken scale or other measuring device – you can get the very best readings you can, but if the scale is off, it will never be the 'right' number (the accepted value) for the item you are weighing.

Finally, I talk about me – I threw the ball 100 times, and 80 of them we had to chase the ball out into that empty field. I try to be a little funny about it, then we evaluate. Was I accurate? Not even close. Was I precise? Not at all – I was all over the place. After we hash that out, we are ready to let them try.

Activity – group discussion. Now, I have my students ability grouped (or ability scattered as I like to call it,) so that each team of four is a low, a medium-low, a medium-high and a high student (see my notes about teams if you are interested in this technique). The cards for this activity are color coded so the easy one (both accurate and precise) goes to the low seat, and the hardest one – neither – goes to the high seat. The activity will work without the extra planning but giving each student a question they can answer topic they can speak to helps build their confidence and encourages them to participate in such activities through the year.

I have them talk about their cards for a minute or two (it won't take long), then we come back and big group talk about the results. Finally, I explain that these targets represent the numbers we will be getting when we take our measurements – the bull's eye being the book value (accepted value), and the darts being each of the measurements that we take.

Activity – notes writing. I have the students draw their own targets and label them in their notes, or you can give them preprinted targets to glue in and label correctly.

Page of Targets – print and cut, then glue on cards and laminate for durability.

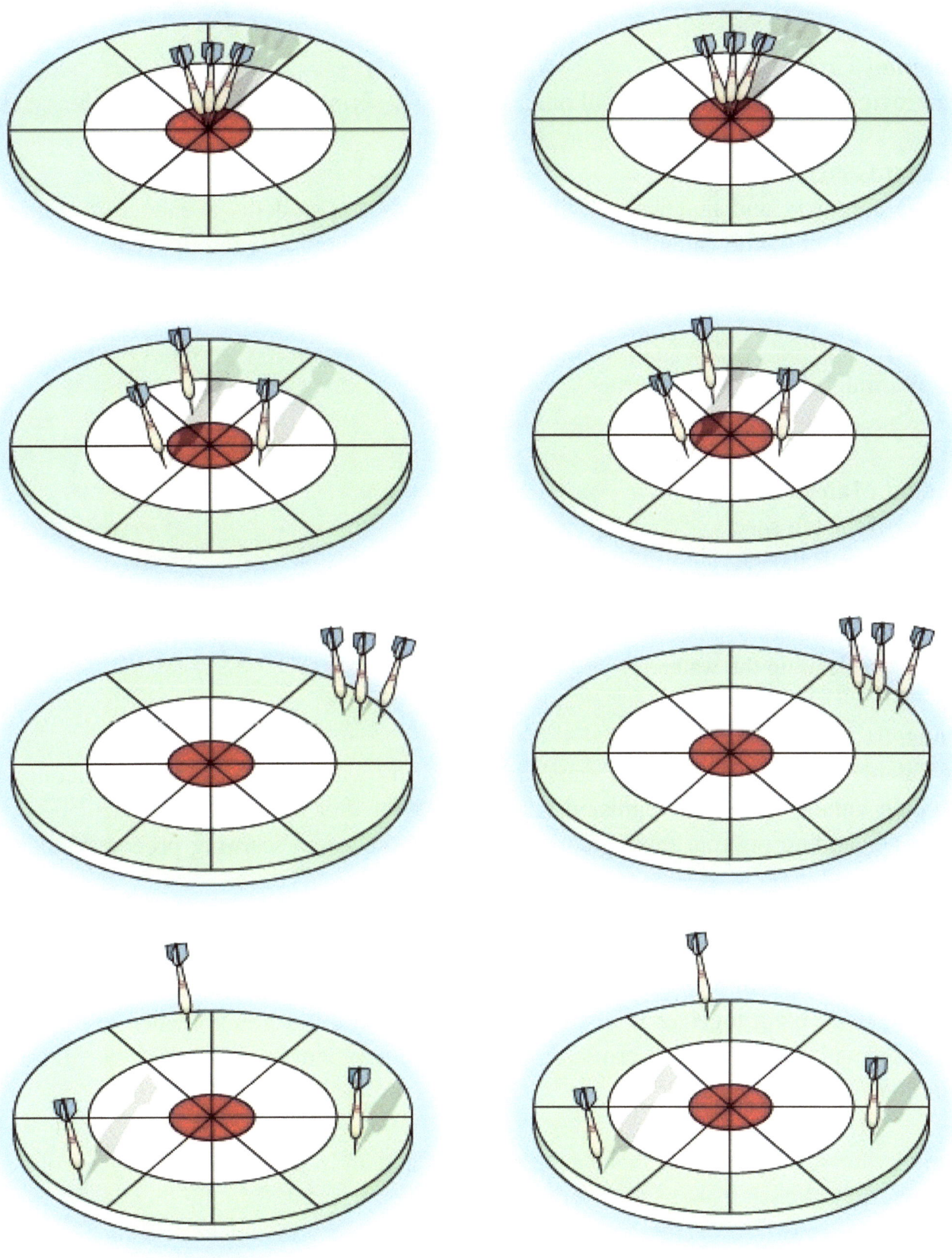

Topic: Soda Lab
Day: 5
Unit: Properties of Matter

Learning Target:
I can review calculating density and other property descriptors with the Soda Lab.

Student Goals:
This is a review and practice activity, which gives the students a chance to name some properties. Along with calculating density, which they should be familiar with, you can also introduce **percent error.**

Agenda:
A – Warmup: Review of Properties
L/A – Soda Lab

Student Materials:
Copies of the lab sheet
Triple Beam Balances
Large graduated cylinders
A variety of sodas
Towels to clean up the water

Props:
Pre-calculate the density of your cans of soda if you want exact numbers. Generally, I use accepted values of 1.1 for regular drinks and .95 for diet drinks, as most will be close to these. Depending on your equipment, this lab is more about learning procedure than being accurate.

Actions and Rationale:
Warmup – review of properties. This lab has lots of steps and brings in several aspects from the unit, tying them together in a single activity. It also gives you a chance to observe your student's math skills and their proficiency with the equipment, which will be useful for future planning.

Begin by having the students make a Q&Q t-chart and an I&E t-chart for their can of soda – I give one can to each team of four, so they will all be different. Remind them to be specific – give properties of their actual can, not generic terms, but you can remind them of some of those generic terms to get them started, such as saying **300 mL** rather than **volume**.

Lecture – lab procedures. Go through the lab and demonstrate each step – I try to never assume they know how to use the equipment, as there will always be students who do not.

1. Using the scale to get the mass – have them do this before they get the sample wet for the best reading. Be sure to go over each step and how to read it.
2. Using the graduated cylinder to get the volume – we have 1000 mL plastic ones. If you don't have them, they can use the volume on the can and adjust your accepted values accordingly. Go over how to use it, including getting down at eye level to read it.
3. Calculate the density – which is mass divided by volume. This is a good time to talk about how density works, and that objects that are less dense float on top of items that are denser. Did their can sink or float? This is a good indicator of how their density should have turned out since the density of water is 1 g/mL It's also a good time to point out that density is intensive – the density of each can for each type will be the same.
4. Percent Error – give the students the accepted value for their can and ask them to calculate their percent error.

Activity – the Soda Lab. Due to time, I have my students get their measurements and go as far as they can with finding their answers, but if they have to take it and finish at home, they can do that as well. If you have less time or want to be sure they finish, you can skip the percent error calculation, or you can do the lab together as a class using a single can of soda and then do their calculations individually. For me, part of the objective is for them to experience the hands on of the activity, so scale and modify as needed.

Soda Lab

1. Brand of Soda _____

2. Mass of Soda _____

3. Volume of Soda _____

4. Density of Soda _____

5. Accepted value of soda _____

6. Percent Error _____

    Calculations:

$$D = \frac{m}{v}$$

Percent Error = $\frac{|\text{accepted value} - \text{experimental value}|}{\text{accepted value}} \times 100$

Conclusions and Reflections
1. Based on the density, will your soda FLOAT or SINK in water? _____
2. Why do you think so?
   _____
   _____

3. What does your percent error tell you about your results?
   _____
   _____

© Lavish Publishing, LLC

Topic: Reading Day
Day: 6
Unit: Properties of Matter

Learning Target:
I can read and annotate a science text, then answer questions about how matter particles act in each state.

Student Goals:
Students understand how the states of matter can be described in terms of volume, shape, compressibility, and movement of particles.

Agenda:
A – Warmup: copy notes
A – Article Lesson: Properties of Matter

Student Materials:
Journals for Notes
Copies of articles for students

Props:
Chart for students to copy

	solid	liquid	gas
Volume	Definite	Definite	Indefinite
Shape	Definite	Indefinite	Indefinite
Compressibility	Not compressible	Not compressible	Compressible
Movement of particles	Vibration only	Sliding motion	Full movement

Actions and Rationale:

Warmup – copy notes. The students copy the chart into their journal as a quick review of how the molecules in a sample interact depending on the state of that matter. You can then give them a quick overview of it, or depending on the article you chose, you can simply turn them over to the activity.

Activity – Students read and annotate using the strategy that they learned during the first unit. Then they answer and submit their questions.

Article suggestions – Our textbook has an article that I use for this, which provides the questions for me. However, since you might have to look elsewhere, there are some good articles on the web about how sugar decomposes rather than melts. You can make up your own questions for this, or I have had the students write their own persuasive paragraph in lieu of questions – is melting sugar chemical or physical and defend your opinion. And of course, you can keep it simple and find an article that describes the relationships in detail and again, make up your own questions or have them write a reflection paragraph on it.

Topic: Matter Stations Lab Activity
Day: 7 - 9
Unit: Properties of Matter

Learning Target:
I can explore a variety of physical and chemical changes and properties through a Matter Stations Lab Activity.

Student Goals:
Students will get an overview of what is covered through the year – includes reactivity, solubility, molecular interactions, and visual properties.

Agenda:
A – Warmup: Change Paragraph
L/A – Matter Stations Lab

Student Materials:
Slips of paper for the warmup
Copies of Lab Sheets

Props:
Lab Set Up – stations with direction cards and materials (9 total with 2 optional) copied, laminated, and tables prepared in advance

Actions and Rationale:
Warmup – change paragraph. Have students write six to eight sentences about the differences between physical and chemical change. I take them up for points and we transition into the lab.

Lecture / Activity – matter stations lab. Walk the students through the stations and your expectations at each. This lab is slated for three days so you can be thorough covering this (fifteen minutes or so) and they can take their time at each station. Below is what I tell my kids for each station. Then there are prep instructions for each station, and finally the lab report and table instruction cards. This lab takes a lot of set-up (a couple of hours) so don't rush it inside the classroom. If you don't mind setting up a different chunk each day – two to three stations for each – then breaking the lab up into smaller chunks is acceptable, but I find putting it all out and just rotating works best – and I only have to set up each station once. After the lab is over, I move them to a side counter where those who were absent can come in and make up whatever they have missed.

Physical Properties Student Instructions – use the magnifying glasses and look close at the structures of the samples. They are all a shade of 'white' so you might want to go with shades. As part of the luster – are they bright, or dull, or clear looking. Describe texture by how the sample particles interact… clumpy, smooth, soft, etc. For structures, you might put little balls, like sand, or even like little boxes or cubes for crystalline.

Setup Directions – I use black construction paper and label each of them with the common name, chemical name and the chemical formula where appropriate, as these could be used for an extension lesson as well. You can also set a small cup of each item on the table with them (I use the small disposable bathroom cups with the name on the side). The items we explore are: sugar / sucrose ($C_{12}H_{22}O_{11}$), table salt / sodium chloride (NaCl), Epsom salt / magnesium sulfate ($MgSO_4$), talc / magnesium silicate mineral ($Mg_3Si_4O_{10}(OH)_2$), baking soda / sodium bicarbonate ($NaHCO_3$), and corn starch ($C_{27}H_{48}O_{20}$), which doesn't have a chemical name we would want to write. Also note, I use baby powder for my talc.

Place the paper with the substances logically on your lab table so a few students at a time can visit, compare, and talk about what they are seeing. The area needs to be well lighted. Provide small hand lenses for each set for close examination. There should also be room between them for the students to place their lab report while writing their observations.

Condensing and Boiling Student Instructions – these are LOOK but DON'T TOUCH! I say, "The hot plate is…" and let them fill in the word **hot**. Then I say, "So don't touch it. Look and observe but leave it alone." Same with the condensing beaker – I warn them not to touch it or it messes up the sample, which is collecting under the watch glass. I squat down and show them how to look under at eye level – we will do this many times this year, so they need to learn and use eye-level point of view whenever needed.

Setup Directions – this station will need a water source, as the pan will have to be refilled as the water boils away. You will also need ice nearby, as the beaker will need to be freshened about once an hour if you have multiple classes using it.

I use a plain saucepan for the vaporization – I have a couple I keep in my room for such activities. I also position this table next to the **viscosity station** so that I can pour the hot water into the 'hot' beaker when needed (usually I reset between each class). You will need a hotplate to place the pan on, and an area the students can write on nearby that will have plenty of clearance to avoid contact with it.

For the condensation beaker – any size will do, but I like to use a 250 mL or so, with a watch glass to rest on the top of it. Place a small amount of water in the bottom of the beaker – maybe half an inch and set it on the table away from the edge where it won't get bumped by accident. The watch glass goes on top so it forms a shallow bowl, and you add a few pieces of ice – but not too much or it will leak over as it melts. The condensation will collect on the underside of the watch glass, so they really do need to squat down and look through the side of the beaker for the best view. And again, give the students some room for observing, sharing, and writing that will not disturb the set.

Viscosity Station Student Instructions – be careful with the tubes. If you drop them, they will break! Pull them out one set at a time, turning them upside down so the little balls inside can "race" each other. Observe which move faster and which move slower – be sure to compare pancake to corn – hots. Then compare corn to corn, hot and room. Finally, compare the different temps of pancake to each other. Be sure to record what you observed on your lab sheet.

Setup Directions – this station is a bit more complicated, so we will look at it in steps.

Step one – setup your viscosity tubes. I bought mine from amazon cheaply – glass test tubes with screw on tops. You will need six but having extras won't hurt. Build the tubes by filling three of them with corn syrup and three with pancake syrup. Each tube also needs a 'ball' of some sort – tiny marble, lead shot, etc. I used bbs once, but they corroded in storage.

Step two – set up the containers. You will need three beakers that the tubs fit in with a little room around them – mine were 400 mL. Fold a paper towel and place it in the bottom of each to create a little cushion between the tubes and the beakers. One tube, you label as **Ice**, the second is **Room**, and the third is **Hot**. I like to write on mine with a sharpie – which will clean of with a dab of rubbing alcohol later.

Step three – set up the station. You will pack the ice beaker with ice, then fill it with water and work the tubes in. It will keep for about an hour before the ice is all gone. The warm beaker stays dry, so just set the tubes in it. The hot beaker will also need to be refreshed, which is where the pan of hot water on the next station comes in. To start the day, I heat a pan and fill the hot beaker. After that, I dump the beaker and refill it between each classes to keep it fresh. DO NOT put the hot beaker on a hotplate – overheating the tubs will cause them to explode when the pressure becomes too great.

Solubility Station Student Directions – again, be aware of the glass and handle with care. The lab says to use about ten mL of **distilled** water. I hold up a wash bottle and remind them – "This is NOT for city water. If you run out, let me know and I will refill them properly. Please do not refill them from the sink." Then I explain that the measurement does not have to be exact. Instead, we will guestimate the water using our finger against the side – two knuckles worth (joints of the finger) is about right.

Then I hold up a scoopula with just the tip full of powder and tell them, "Use your solute sparingly. If you put too much, you won't be able to tell if it is dissolving, so go with just a little." I also point out that each cup has a scoopula for it – remind them not to mix them around or it can cause contamination and ruin their results.

I show them how to pour in their solute and insert the stopper. Remind them not to use their fingers – that is a bad habit that could have consequences when we use 'real' chemicals. They should shake them long enough to allow the solute to dissolve if it is going to.

Then the tubes have to be 'rested' in the test tube rack. They will need a marking strategy so they know which solute went into which tube, perhaps a template that lays in front of the rack and points at which hole should be each tube / solute. Waiting can be the hardest part, so maybe setting a 1 minute timer or watching the clock will help.

Finally, we talk about what it will look like after the tubes set for a minute or so after shaking. If the solute collects on the bottom of the test tube, it was insoluble. If it 'disappeared' then it was soluble. It's not gone – it is just dissolved, like sugar in our tea when we like it sweet.

This table will also require some **cleanup**. I keep old bath towels in my room for such things, and there will be one at this table to wipe up drips of WATER. We don't use the towels on chemicals, but we have lots of water spills, and they are perfect for that – and I have a rack in the back to hang them on for drying after the lab. They also need to wash out their tubes, so they need access to a sink and a test tube brush – demonstrate how you want them to leave the station before they move on.

Setup Directions – this station will need four test tubes, four rubber stoppers, a test tube rack with solute template, four cups with each of the solutes, a marked scoopula for each cup, and a plastic wash bottle of distilled water. The solutes being tested are: sugar, table salt, Epsom salt, and talc. Label the cups and the scoopulas with a sharpie – remember that it will come off of the scoopulas with alcohol when you are ready to clean them. Paper cups I toss when we are all finished, but plastic ones I empty into the trash and wash to use again – we will be using these materials numerous times through the year.

Precipitate Station Student Directions – this station happens in two parts. First the students will make solutions from each of the four solutes just as they did at the solubility station, so I suggest they visit that station first. Then they can come to the precipitate station knowing how to set up their tubes, only this time they won't empty them right away.

Instead, they will use Silver Nitrate to perform a test and see if a chemical reaction indicator forms – a precipitate. Show them how to use the dropper bottle to apply the drops – and warn them that it will make black spots on their skin when exposed, so if they get any on them, they will need to wash their hands with soap and water – a great time to point out that germex is not the same as washing. Silver Nitrate can also stain their clothes, so they need to be careful while using it.

For the results, a precipitate will be a thick white blob that either floats or sinks to the bottom. If they gently agitate the tubes, it may help with forming a good precipitate, but not too harshly. Also, if the result is negative, adding more silver nitrate will NOT change the result – it is simply being wasteful, so I strongly caution my students against it.

Once they have finished their testing, they will need to clean this table in a similar fashion as the solution table. Be specific here if you have any other directives so they know what to do.

Setup Directions – this table is exactly like the solutions table, except for different solutes and the addition of the dropper bottle of **silver nitrate** – make sure it is labeled. A .5 molar solution is fine – it doesn't have to be exact to work. Simply make or purchase the silver nitrate and test it for yourself so you have an idea of what you are looking for on lab day.

The **solutes** we are testing are: sugar, table salt, Epsom salt, and baking soda. Again, label everything, including the slots for the tubes, the cups, and the scoopulas.

Reactivity Station Student Directions – these directions will depend on the tray option that you choose. You will want to hold it up and show the students a small amount on the tip of your scoopula again – tell them that large amounts of reactants do not change the result and are only wasteful. Overdoing it will also make the cleanup harder, so they should also keep that in mind.

Tell your students to set up all the powders first. After their reactivity tray is set, they take the water and apply it all the way down the column. Then the iodine, and finally the acetic acid (vinegar). Remind them that one or two drops is sufficient to see a result, or the lack of one. If it doesn't react, adding more will not change that fact.

Reading the results – water is the control, so there will be no reactions with it, but it gives the students an idea of how a negative result will look. The iodine will react with your starches, so it may react with your baby powder if it has starch in it, and of course the corn starch – the color change is black or dark violet / blue. The vinegar will bubble or sizzle in the baking soda.

Cleanup on this one will depend on the trays you use. I always have my students dump their tray into the trash, so there will need to be access to a trash can or other receptacle, like a bucket. If they are using actual reactivity trays, they will need to be rinsed WELL in a sink and left clean for the next group. If they are using the laminated trays you built, they can be wiped down with a damp cloth. If you choose paper plates, the whole thing can be thrown away and new plates put out for the next group.

Setup Directions – you will need your sample cups set up as they are on all the other tables, labeled with scoopulas. For this table, you will be testing baking soda, table salt, Epsom salt, talc, and corn starch. You will be checking for reactions with water, iodine, and vinegar, so you will need small dropper bottles for the iodine, and I use small squirt bottles for the vinegar and water. You can use a cup for them with a pipette, but that leads to possible contamination.

For the **reactivity trays**, drawing them onto **paper plates** ahead of time requires more setup ahead of time, but it is the fastest and easiest cleanup for the students during the lab, even if it is more wasteful. You can use **real tray**, which come in different styles, but I found that if the wells were deep (which mine were), the students felt like they needed to fill them completely, which led to little compartments caked with powders that didn't wash out easily, and you know kids… it was a hassle. Plus, if you use real plates, you will need to have a template that they can set the tray on that has the items labeled down the side and across the top, or they will get mixed up on which was what.

A few years ago, I made some **fake plates** with black construction paper. I make a template for the tray complete with labels on the sides and across the top for each of the reactants, a simple grid with a circle in the center of each 'well' to mark where to put their powder. These were very successful, with only those totally bent on using extra making a mess, which wasn't that often. To clean them, the wiped them of in the trash and then used a damp paper towel to clean the top surface and leave for the next group.

Sample Bottle Station Directions – this station has several bottles that are sealed. I direct the students – "We NEVER open sealed containers unless you are told to do so!" I use the magnet and demonstrate how to use it through the plastic bottle – do it on the copper, since it won't be attracted to the magnet. Tell them to try the magnet on all of them to see if anything is attracted to it.

Setup Directions – each bottle will need to be prepped ahead of time, then placed on the lab station with a magnet and room for the students to talk and write their observations. This is a dry station, so I use a table with chairs that they can sit in and be comfortable for a few minutes while they complete this one. For the bottles, I use old water bottles and for each you will need…

Copper – these can be fillings or pellets.

Aluminum – again, this can be pellets or some other form.

Brass BBs – I bought a little container for this and other items – a nice amount works.

Iron and Sulfur – do NOT use actual sulfur. I know we warn them not to open the bottles, but sometimes they don't follow directions, so it is better to be safe. I got a box of yellow chalk for this and crushed up a few pieces in a mortar and pestle – it looks great and is a perfect substitute. Add your iron filings and you are good to go.

Copper II Sulfate Solution – again do NOT use actual $CuSO_4$ in case they open the bottle. Instead, put a few drops of blue food coloring in plain water, and you are good to go with a reasonable facsimile.

Density Station Student Directions – another dry station if you use cubes, or wet if you use graduated cylinders, so your directions will depend on what you have and how you set it up.

I have the students calculate their density on a sample, which you can have a selection to choose from or have them all do the same one. Be sure to give them step by step on the scale – some of them will not know how to use a triple beam balance or even a graduated cylinder properly. If they are using a cube sample, give them a ruler and review how to calculate volume given the dimensions.

After they calculate, they use a density table for the elements to decide which element they have based on the density and the physical properties they can see. Finally, I have them check if they are correct (this can be at another location, or they can see you to verify their choice). They take the value of the CORRECT metal and use that as an accepted value and theirs as an experimental value to calculate their percent error.

Setup Directions – you can choose the best way for your classroom. I am out of lab stations by this point, so I use cubes and a ruler for the volume, and of course a triple beam balance for the mass. I set it up on a folding table with room to write on and put a list of metal densities for them and calculators as well.

If you have room and want to use graduated cylinders, I recommend plastic ones – if they are using metal ones, it increases the odds one will be broken. If you use glass, be sure to show them how the glass guard works – a large number of them think that sliding ring is for measuring.

After they measure and calculate, they visit with me to verify they found the right metal. They use the actual density to calculate their percent error.

A note about **calculators**. I do not let the students use their phones as a calculator in my class. I make them use one from the set that I provide – all the same. They practice during class time before hand, and this is the only one they get on the test. This is a good habit and I suggest it to everyone whenever possible.

Flammability Station Student Directions – this station MUST be set up under a fume hood. If you don't have one available, do it as a demo well away from students in a well-ventilated area, but I never let students do this type of test on lab tables. If you do, do not use the alcohol – that way they are testing items that are NOT flammable. Call me extremely cautious, but to me it is worth knowing they will be safer.

When explaining it to the students, let them know that when you are flame testing anything, always err on the side of caution – full protective gear, fume hood, hair and clothing lab appropriate. You will also want to be close to this station at all times just to monitor what they are doing – some kids like to get creative, and that's one thing that makes this station so dangerous.

For the test directions, show the students the wood splint and the evaporating dish. They will need to put a SMALL amount of item to be tested in the dish – five or six drops, again more won't make the test better, and will make cleanup more difficult. They will NOT be holding the dishes when they apply the flame – show them how to set them on the name cards you placed inside the fume hood ahead of time.

Explain that they will light the Bunsen burner inside the hood and use it to light their wooden splint before they apply it for the test. This is a great time for you to demonstrate how to properly light one, and for them to practice their lighting technique. I also warn them that the striker is not a toy – please do not strike it repeatedly for no reason.

When they are ready to test, they light their splint. They place the splint close to the dish and touch it to the sample. If the item is flammable it will be destroyed, as it is going to catch on fire and burn. If it is not flammable, it won't.

After the test, the students need to blow out and put the burnt splints in the 'burn can' which is also inside the fume hood – I use a metal coffee can for this. Check it periodically to make sure nothing inside of it is smoldering. The dishes will need to be washed and dried, returned to new condition for the next group to come to the station.

Setup Directions – you will need squirt bottles or containers of the four items you are testing, which are water, vinegar, rubbing alcohol, and pancake syrup. I put the bottles in an adjacent space or counter where the students can set up their dishes and then move to the flame testing away from them.

Under the hood, you will need a striker, Bunsen burner, wooden splints, and your waste can. Again, this station will require the most thought, preparation, and monitoring, so it is the first one that I cut if my classes are large, unruly, or if we have time constraints.

Properties & Reactions Station Lab

Condensing & Boiling - are these processes physical or chemical and how do you know?

Viscosity - how did the temperature affect the outcome?

Precipitate formed? (Y or N)

Sugar _____

Table Salt _____

Epsom Salt _____

Baking Soda _____

Are precipitates physical or chemical? how do you know?

Visual Properties - describe what you see

	color	structure	luster	texture
Sugar				
Table Salt				
Epsom Salt				
Talc				
Corn Starch				
Baking Soda				

© Lavish Publishing, LLC

Reactivity with Reagents (explain what happened and what it means)

	Water	Iodine	Vinegar
Baking Soda			
Table Salt			
Epsom Salt			
Talc			
Corn Starch			

Matter Sample Bottles - Mixture or Pure Substance, why?

	Mixture	Pure Substance	Reason
Copper			
Aluminum			
Brass BBs			
Iron and Sulfur			
$CuSO_4$ Solution			

Soluble in Water (Y or N)

Sugar	___
Table Salt	___
Epson Salt	___
Talc	___

Is solubility physical or chemical? How do you know?

© Lavish Publishing, LLC

Flammable? Y or N

Water _____

Vinegar _____

Alcohol _____

Syrup _____

What happens to the sample when the test is negative? When it is positive?

Density Station

Density is _____

Metal is _____

Calculate the density of you sample - show your work

Actual Metal _____

Percent Error _____

Calculate your percent error - show your work:

© Lavish Publishing, LLC

Condensing & Boiling Station

DO NOT TOUCH –
this station is for observation only.

Here you see a pan of boiling water. Notice the steam coming off. This process is called _____.

You also see a beaker with a watch glass on top. There is probably a piece of ice sitting on top of this, cooling the interior of the beaker. Be sure to get down to eye level to look at the UNDER side of the watch glass. This process is called _____.

On your paper, decide if these processes are physical or chemical and give two reasons for each that explain how you know.

CLEAN UP – leave the station neat and ready for the next group.

© Lavish Publishing, LLC

Viscosity Station

Here you see a selection of sealed glass test tubes. Be careful not to drop, bump or break them.

Some are in cold ice water, some at room temperature, and some in warm / hot water. Gently lift one tube at a time and turn it so the ball inside slides from end to end. Note how long it takes for the ball to make the trip. This is one way to measure viscosity, which is the resistance to flow. Flow is how FAST a liquid moves from one place to another, so viscosity is the OPPOSITE of flow, or how slowly it moves.

Viscosity is a _____ property. How did the temperature of the tube affect the viscosity?

CLEAN UP – return the tubes to the correct location – do NOT switch them from hot to cold or cold to hot. Mop up water drips with a towel and leave ready for the next group, please.

© Lavish Publishing, LLC

Precipitate Station

This is a TWO PART station.

First, make small amounts of solution for each:
1. put 10 mL or so of distilled H_2O in a test tube
2. add a small amount of one of the solutes
3. put the rubber stopper in the top and shake the tube gently to dissolve.

Second, test for reactivity with silver nitrate:
DO NOT GET THIS ON YOU – it will stain!
1. put the tubes in the rack without the stoppers
2. Add a few drops of silver nitrate to each and observe – cloudy or clear is negative, bits of white solid that float or sink to the bottom are precipitate
3. record your results and answer the question.

CLEAN UP - Empty all of your test tubes in the sink and rinse them and the stoppers. USE WATER ON LOW or it will splash everywhere. Leave ready for the next group, please.

© Lavish Publishing, LLC

Visual Properties Station

Feel free to TOUCH, but please do not make a mess

First, look at the material on the paper. Then, use the magnifying glass to get a closer look. Record your observations on the grid.

Color – these are all shades of white, but there are subtle differences you might note.

Structure – is it crystalline, powdery, or something else.

Luster – is it bright, shiny or dull.

Texture – smooth, rough, slick, slippery, clumpy, sandy... how does it feel?

CLEAN UP – leave the items as you found them – neat and ready for the next group, please.

© Lavish Publishing, LLC

Reactivity Station

This is a TWO PART station.

First, set up your reactivity tray with a SMALL amount of each substance in the boxes for that row. Do all of them before you move on.

Second, add ONE or TWO drops of each reagent to the small amount of powder that you are testing. Record your results.

Reading the Tests:
Water (the Control) will not react, it will be wet.
Iodine turns blue or purple in the presence of a starch. Negative results will not change color.
Vinegar (acetic acid) will react by foaming with some substance. Negative results will not.

CLEAN UP – Gently dump your tray into the trash and clean it as directed. Leave it ready for the next group, please.

© Lavish Publishing, LLC

Matter Sample Bottles Station

Do NOT open the bottles or destroy the labels

Look at all five bottles.

If there is only ONE substance in the bottle, we call this PURE. Pure substances can be elements or compounds, but they are all the same material.

If there is more than one, it is a mixture. Mixtures come in two types and can be separated by physical means, and there are lots of those!

On your lab report, record your thoughts and reasons for each.

CLEAN UP – leave the station neat and ready for the next group.

© Lavish Publishing, LLC

Solubility Station

Follow the steps:

1. put 10 mL or so of distilled H_2O in a test tube
2. add a small amount of the solute
3. put the rubber stopper in and shake the tube gently to dissolve the solute.
4. Allow the test tube to sit for a minute.

Make observations - Did some or all of the powder (solute) disappear? Look at the bottom to find out.

If so, then that item is _____ – mark Y on your lab report. If not, it is NOT _____ – mark no. Be sure to answer the question there.

CLEAN UP - Empty all of your test tubes in the sink and rinse them and the stoppers. USE WATER ON LOW or it will splash everywhere. Clean up any messes to leave ready for the next group, please.

© Lavish Publishing, LLC

Density Station

Cube Shaped Samples

You have a cube made of some metal (element) from the periodic table. Use the ruler to measure the side (or sides if it is rectangular) and determine the volume of your cube. Remember: V= l x w x h.

Use the triple beam balance to obtain the mass of your cube.

Finally, calculate the density of the cube. Using your chart of element densities, which metal is it?

Obtain the actual (accepted value) density of the metal from the designated location and calculate your percent error.

CLEAN UP – leave the station neat and ready for the next group.

© Lavish Publishing, LLC

Density Station
Oddly Shaped Sample

First - Use the triple beam balance to obtain the mass of your sample. If you use a weigh boat, be sure to subtract it from the mass of your sample.

Second - Use the graduated cylinder to measure the volume. Use the water on low to avoid splashes. You will need the initial volume of water, add the sample and read the final volume. Finally, subtract (final minus initial) to get the volume.

Third - calculate the density of the sample. Using your chart of element densities, which metal is it?

Obtain the actual (accepted value) density of the metal and calculate your percent error.

CLEAN UP – Pour the sample through the strainer so you don't lose any pieces. Dry them with a paper towel and leave the station neat and ready for the next group, please.

© Lavish Publishing, LLC

Flammability Station

WARNING – Safety googles and aprons required!

Set up – place 4 to 5 drops of a substance to be tested in an evaporating dish. When all the samples are in their dishes, move to the fume hood. You will be using FIRE – so use extreme caution at all times!

Under the fume hood:

1. Place your samples in the center on the name cards.
2. Light your Bunsen burner and adjust it to a short yellow flame.
3. Take a wooden splint and light it with the Bunsen burner.
4. Apply the flame to each of the samples – observe if they catch fire or not.
5. Blow out the wooden splint and place it in the burn trash can.
6. Turn off the gas to the Bunsen burner.
7. Move your samples to the sink – wash the evaporating dishes completely and leave them clean and ready for the next group.

USE WATER ON LOW or it will splash everywhere. Clean up any messes to leave ready for the next group, please.

© Lavish Publishing, LLC

Topic: Element or Compound
Day: 10
Unit: Properties of Matter

Learning Target:
I can review and practice the difference between elements and compounds with a sorting activity.

Student Goals:
Students will be tested on knowing the difference between pure substances as elements or compounds, including picture examples. These terms should both be familiar to them, as this is largely a review activity and a setup for what comes next.

Agenda:
A – Warmup: Discussion Pictures
L – Elements and Compounds
A – Partner / Group Discussion Cards

Student Materials:
Journals for notes
Sample Picture Cards (of actual items)
Answer Documents from Matter Stations Lab

Props:
Sample Picture Cards (enlarged drawings)
Sample bottles from the Matter Stations Lab

Actions and Rationale:
Warmup – discussion pictures. Post the Sample Picture Cards 1 - 4 under a doc-cam or make copies that students can pick up on their way in to look at. I keep a class set of the full page on colored paper and laminated so we can use them several times and simply direct them to use 1 – 4 for the warmup. They will need to observe these four and decide if they are an element or a compound based on their prior knowledge: You can have them record their responses, share their thoughts with their partner, or simply hold them in their mind for the lesson.

Lecture – elements and compounds From the warmup, we transition into our discussion of what elements and compounds look like and the picture cards are the perfect up-close look at them. It also gives us a chance to talk about some things that we will come back to time and again through the year.

Most of them should recognize picture one as an element. Point out that we call them monatomic because MONO means one, and there are single atoms floating around in there.

Then I skip to picture three, which is a compound, and I say, "This could be C_2H_4." From there we discuss how it compares to box one, which is the crux of element versus compound – elements are alone on the periodic table and compounds are two or more DIFFERENT elements chemically combined.

Then we jump to picture two, and I ask, "So, is this an element or a compound?" I give them a minute to talk about it and go from there. Some will say that's an element because they aren't different, but some will want to call it a compound. This is the launching point to talk about diatomic elements and point out that they never go anywhere alone – they always take a buddy. I foreshadow that we will learn more about them later, but for now they just need to know that some elements always come in twos, but most elements are monatomic.

Finally, we come to the last picture, and they should recognize that this is a compound. However, I tell them that this is the most important compound on earth (2/3 of the world) – this is water. I point out how to recognize it (upside down mickey-mouse), and I again foreshadow that we will spend at least three weeks just learning about water. I actually have a large model of a water molecule hanging from the ceiling, so we have used it as a land mark for our drawers under it, but this is our first time to dig in, and I ask them, "Could this be something else?" Of course it could, but the odds of it are slim, so unless we are TOLD that this is something else, we always assume that a compound that looks like this is water.

Activity – discussion cards. Sorting cards can come from a number of sources. I like stock photos like those from a free site where I can pick out examples. These can be handed out when you are ready, or have the students each pick up one or two as they come in. I glued mine to colored paper and laminated them to last longer, and I give them the "please don't roll, bend, fold, chew, write on, or in any way damage my cards" speech at the start of class or when they have them in their hands.

To show them how they work, I use a few LIVE samples from the Matter Stations Lab – the sample bottles containing aluminum and Epsom salt. If they ask about the others, I simply tell them we will have to talk about those later. For today, we have these two. I hold them up and ask, "Which is the element?" Obviously, it's the aluminum because it is alone on the periodic table. Epsom salt, $MgSO_4$, is not – it is a chemical combination of two or more different elements. If you want, this is a good time to introduce chemical formulas, and you can point out just what elements and how many are in it.

Finally, the students can go through their pictures either with their partner or with their group of four and discuss what they have – element or compound – and WHY they think so. While they do this, I walk around and redirect, as it is important that they get into that conversational part by defending their choices.

Matter Sample Drawing Cards

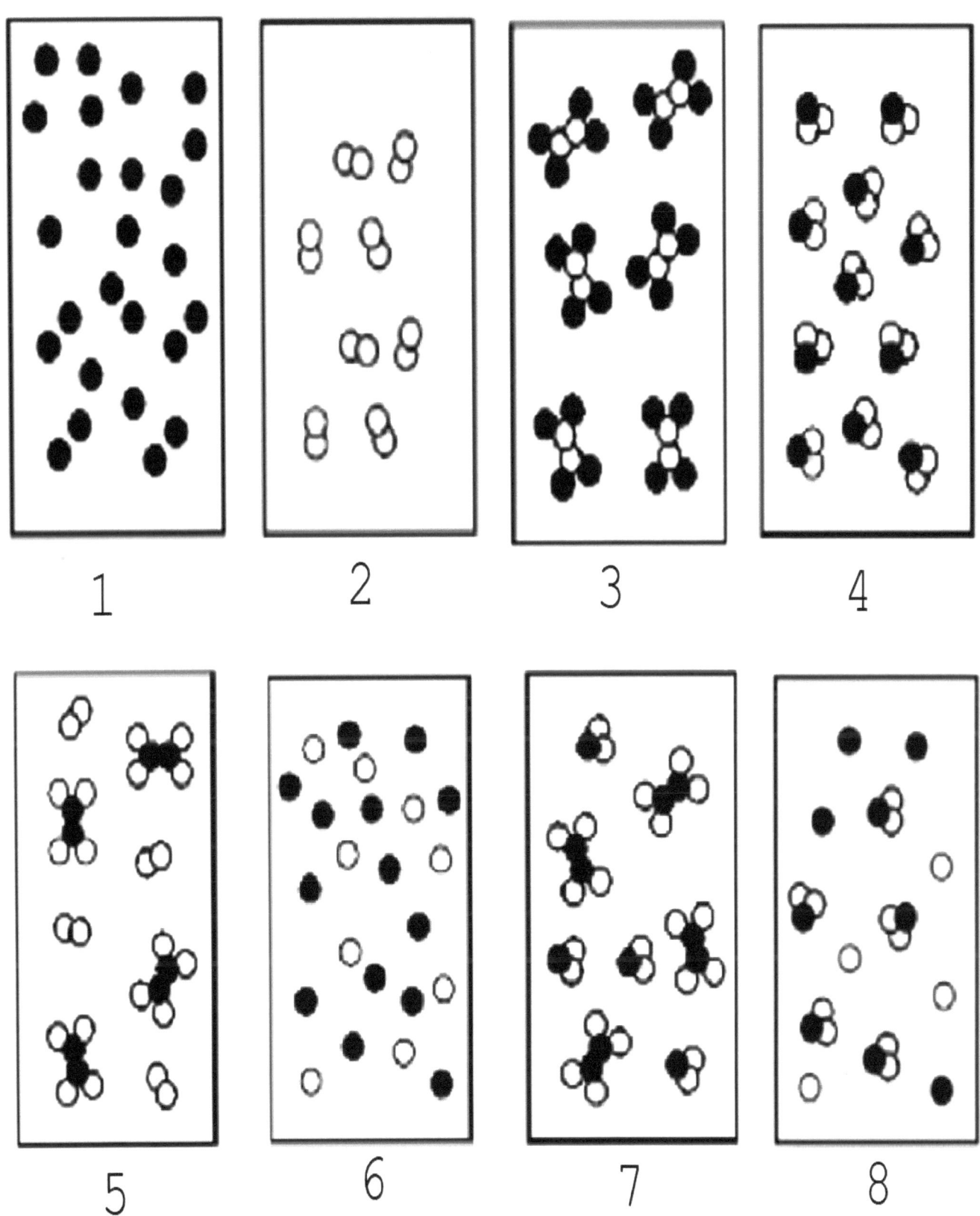

Matter Sample Drawing Cards – answers. Explanations for 1 – 4 are provided with the element and compound lesson. Explanations for 5 – 8 are provided with the homogeneous and heterogenous mixtures lesson. There are further explanations with the mixtures and pure substances lesson for all 8.

These are my interpretations of the drawings and I share with my students that their view of them will grow and change as their understanding grows and changes with expanded types of samples and circumstances.

For our purposes, these are the answers that they want to understand.

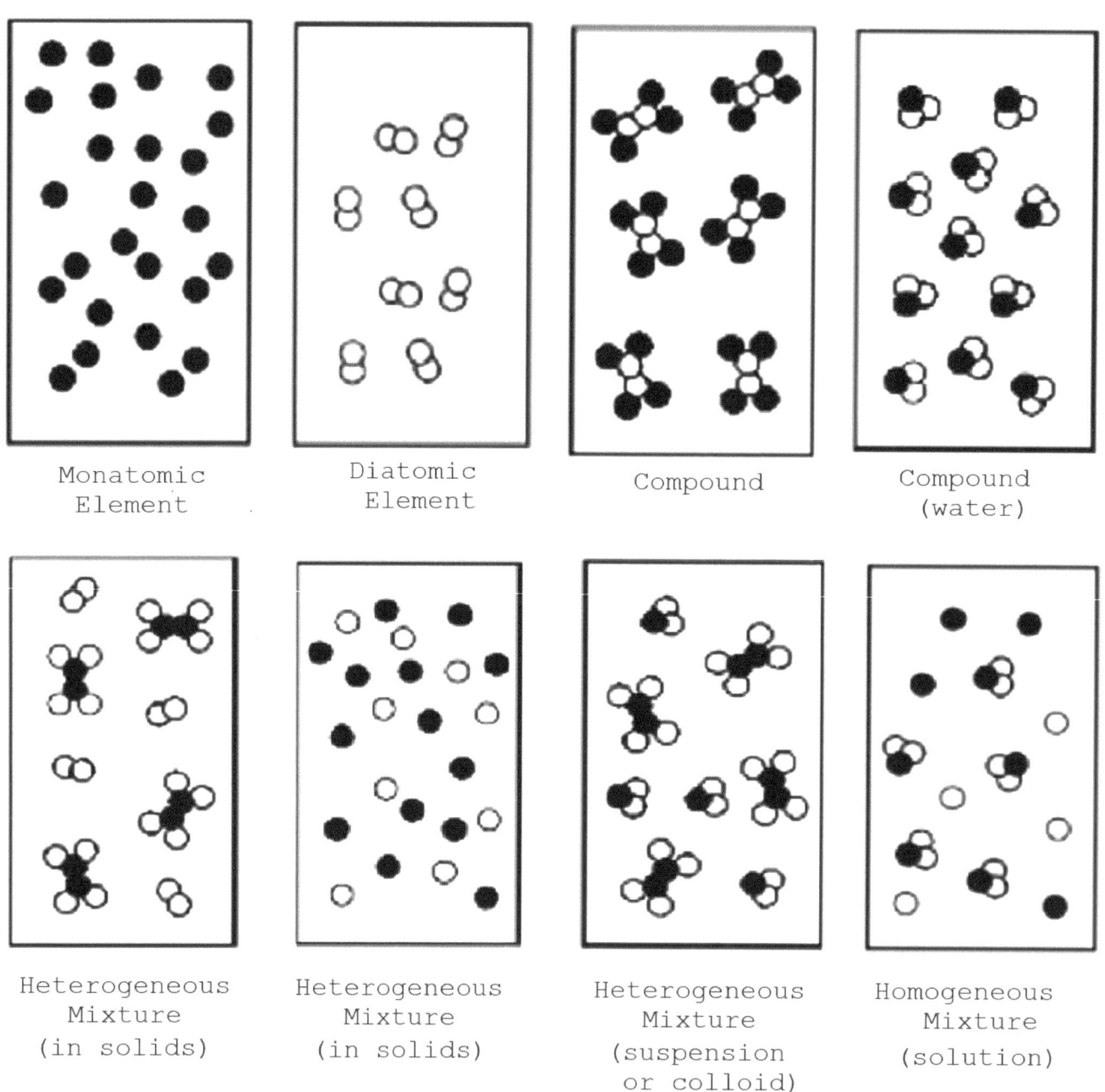

Topic: Pure Substance or Mixture
Day: 11
Unit: Properties of Matter

Learning Target:
I can review the differences between mixtures and pure substances and discuss separation methods using discussion cards.

Student Goals:
Students will be tested on classifying matter as pure substances or mixtures. They should be able to identify that pure substances are all one thing. They should further know that mixtures are two or more substances PHYSICALLY combined, as well as a variety of methods of separating them.

Agenda:
A – Warm Up: Element or Compound Quiz
L – Mixtures and Pure Substances
A – Separation Discussion Cards

Student Materials:
Journals for notes
Slips of paper for quiz
Separation Cards
Answer Documents from the Matter Stations Lab

Props:
Water bottles, label and make one of each... pure water, dirt and water, corn starch and water, saltwater solution, oil and water (I food color the water so we can see it) and table salt. Sample Bottles from the Matter Stations Lab... all of them - aluminum, Epsom salt, brass, Iron and Sulfur, and Copper Sulfate solution.
Separation Word List

Actions and Rationale:
Warmup – element or compound quiz. Have the students complete the quiz while you set up. I take them up for points and then go over the answers as a review from the day before.

Lecture – mixtures and pure substances. You cannot have too many props for this, so I put plenty on the list – once they are made, keep them, as you will be using them time and again throughout the year. You can make them all, or narrow down exactly what you want to use, but I like having options. That way I can steer the discussion and add more where we need it.

Start with the bottle of water. Hold it up and say, "This is all water. What do you think that makes it?" You can refer to the day's learning objective to remind them what the options are – mixture or pure substance. Once you hash out that if it is all the same thing, that makes it PURE, hold up the salt. Then do the salt AND water.

Then I go through as many of the other bottles as we need to – each time it is a mixture I say the AND very loudly. Then we nail it down – if there is an AND in it, it is a mixture of two or more substances. If it all one thing, that is a PURE substance.

Finally, we pull out the sample bottles from the lab and discuss what they chose – did they get them right? If they didn't they should have a better grasp now, but I always double check each, finishing with the brass. Technically speaking, this is a metal compound, but since they are physically combined into an alloy, this is also a mixture. I tell them not to worry, I won't ask them that one on the test.

Activity – separating matter discussion cards. You can do this several ways – I like to make a set of pictures for each group, but if time is short, you can do one for each group that they can share out. You will need to post the list of separation words on the board and provide a way they will find out what each of them is – via dictionary, textbook glossary, or my favorite, google.

As they look up each one (decanting, filtering, hand picking, evaporation, vaporization, distillation, chromatography, magnetism, and sedimentation), they match it to a picture that represents what that process looks like. You could also do these off of a power point after they have looked up and written the definitions, or we share out a few of the more interesting ones from the group. At the end, we top it off with a reminder that all of these processes are PHYSICAL, meaning that the composition of the substances have not been changed.

Here are a few examples of pictures, and again, you can get these at free locations online:

Topic: Heterogeneous and Homogeneous
Day: 12
Unit: Properties of Matter

Learning Target:
I can learn the difference between homogeneous and heterogeneous mixtures and identify examples of them orally with my partner or group.

Student Goals:
Students will be tested on classifying mixtures as **heterogenous** or **homogeneous**. They should recognize that these are PHYSICAL combinations and classify them according to the samples – if they are **varied** or **uniform**.

Agenda:
A – Warm Up: Mixture vs Pure Substance Quiz
L – Sweet Tea and Salad
A – Mixture Sort with Partners or Groups

Student Materials:
Journals for notes
Slips of paper for warmup quiz
Separation Cards (repurposed)

Props:
Salad and Sweet Tea pictures
Mixture Sample bottles from the mixtures and pure substances discussion

Actions and Rationale:
Warmup – mixtures vs. pure substances quiz. I take them up for points and then go over the answers to review the difference between a mixture and a pure substance. Then we transition with the understanding that as pure substances can be split into elements and compounds, mixtures can be divided based on their samples into two categories as well – homogeneous and heterogeneous.

Lecture – sweet tea and salad. Open by pinning down the definition of **varied** and **uniform**. Tell the students that this is the basis for understanding homogeneous and heterogeneous. Have them practice pronouncing the words out loud as a group, that way they get comfortable with our clunky vocabulary – I like to do this when we are learning new 'chemistry' academic vocabulary, so it is a good habit to develop as a teacher.

Once they have a grasp for varied and uniform, I tell them a side story – that I am going to make lunch for them. For the meal, I am going to bring in a giant container of sweet tea and huge bowl of salad. I show them my picture of the tea – if you can find one that has a jug and glass side by side, that is the best, so you can point to them as you are talking. Then I say, "As I am serving everyone a glass of tea out of the same supply, will the samples be uniform or varied?" This question will get mixed results (sizes of glasses and amount of ice confuses some of them), so nail it down – if the samples are coming from the same container and the sugar and tea are all dissolved in there (this is a solution by the way), they are the same. From the first glass to the last glass, the COMPOSITION of everyone's tea will be the same.

I often use a few local restaurants for this, and we talk about who makes the sweetest tea and how it should be the same every time we buy it because the recipe is the same, and therefore the solutions are the same. Then I say, "The same. Uniform. We'll come back to that."

Then we talk about our salad and I hold up a picture of a salad with big chunks of stuff in it. We talk about which parts people might like to get in their salad. If you like lettuce, that is usually right on top. As people are digging around in it, where might we find more carrots? Probably on the bottom. Then I ask, "As I am serving everyone a bowl of salad, excluding size (we already eliminated size for this), will the composition of everyone's salad be the same?" No. Some will have more lettuce, others will have less. Can we pick through the salad and get what we like out of it? Sure – and that means the samples will be varied. It would be impossible to make the bowls uniform, even if we tried – something would always be different.

Ok, time to talk about homogeneous and heterogeneous. I always get giggles at these words, so I go with it. "What does Homo mean?" They laugh, and then I say, "I bet it doesn't mean what you think it means. Not exactly." Then we recall other times they have heard the prefixes – in biology it was homozygous and heterozygous. Many will recognize these. "Homo means 'the same' – remember? Heterozygous means the genes are different, and homozygous means they are identical."

It's the same principle here – if the samples are UNIFORM, they are the same, and that is HOMOgeneous. A few will get the light bulb moment as they realize why we call it homosexual, and I let them have that connection for a second, and then redirect them back to our lunch. "So, which is the homogeneous?" They will likely know that it is the sweet tea. The heterogeneous is the salad, and it all stems from the samples – if they are varied or uniform tells us which kind of mixture it is.

Activity – mixture sort. We will be reusing our separating mixture cards from our previous lesson. Here, they will be able to look again at how things are separated, and that will help give us clues as to what kind of mixtures they are. When you are holding up you bottles for the examples, use the words dissolved and solution often – and point out that those made of homogeneous mixtures, as the first sample is the same as the last one, as in the salt water. When I hold up the dirt water or the cornstarch water, I like to shake them repeatedly and talk about pouring the samples for everyone. They should understand that whoever gets the end of the bottle gets more gunk in their cup. Once you feel good with their responses, have them transition to the pictures.

Again, you can have them pick these up as they come in (remember the rules about handling them), or you can distribute at this point. And of course, this can be a partnered or group activity. While they are working, they can sort the pictures into two groups, homogeneous and heterogeneous, so you can see if they are getting them right. I also like to walk around and hear what they are saying – it keeps them on target, and I can give them hints if they seem to be missing the point or struggling.

Finally, I like to give them a sentence stem for things like this, as it helps to build their conversation skills, especially if they are English language learners. Try something like…

"This is a picture of _____. It is a mixture because it has ____ AND ____ in it. If I pour or make samples of it, they will be _____ (choose varied or uniform), which means it is ____ (choose homogeneous or heterogenous)."

Write the stem on the board and encourage everyone to use it for the day's lesson.

Topic: Matter Flow Chart
Day: 13
Unit: Properties of Matter

Learning Target:
I can combine the concepts of the matter flow chart and use it to identify examples in a station lab activity.

Student Goals:
Students will be tested on the overall flowchart as well as each of the subsections. They need to be able to break down example pictures and live samples.

Agenda:
A – Warm Up: Homogeneous vs. Heterogeneous Quiz
L – Matter Flow Chart
A – Matter Sorting Stations

Student Materials:
Journals for notes
Slips of paper for warmup quiz
Slips of paper for Matter Sorting Stations

Props:
Lecture – sample bottles from the matter stations lab and a few extras in case you need them, at least one of each type (heterogeneous, homogeneous, element and compound).
Lab Set Up – you will need a variety of samples with enough stations (I like 3 per station) for all students to be working at the same time.

Actions and Rationale:
Warmup – homo vs hetero quiz. I take them up for points and go over the answers to review the difference between a homogeneous and heterogeneous mixture. Then we transition into the lecture to learn how to use each of the parts we have learned and build one large flowchart.

Lecture – Matter Flow Chart. If you google this, there are hundreds of them online and most likely a copy in your textbook. The key is to find one that you like or feel comfortable with, as my version is unique. I like to ask my transition questions so that all of the 'yes' responses are on the left of the boxes and all of the 'no' responses are on the right.

This is also the version I use on the exam, so if you go with something else, adjust accordingly. Have students copy the chart in their journal or sketch the outline to fill in as you talk. You can also build this on cards to put up on the board (or power point) one box / question at a time.

Matter Flowchart

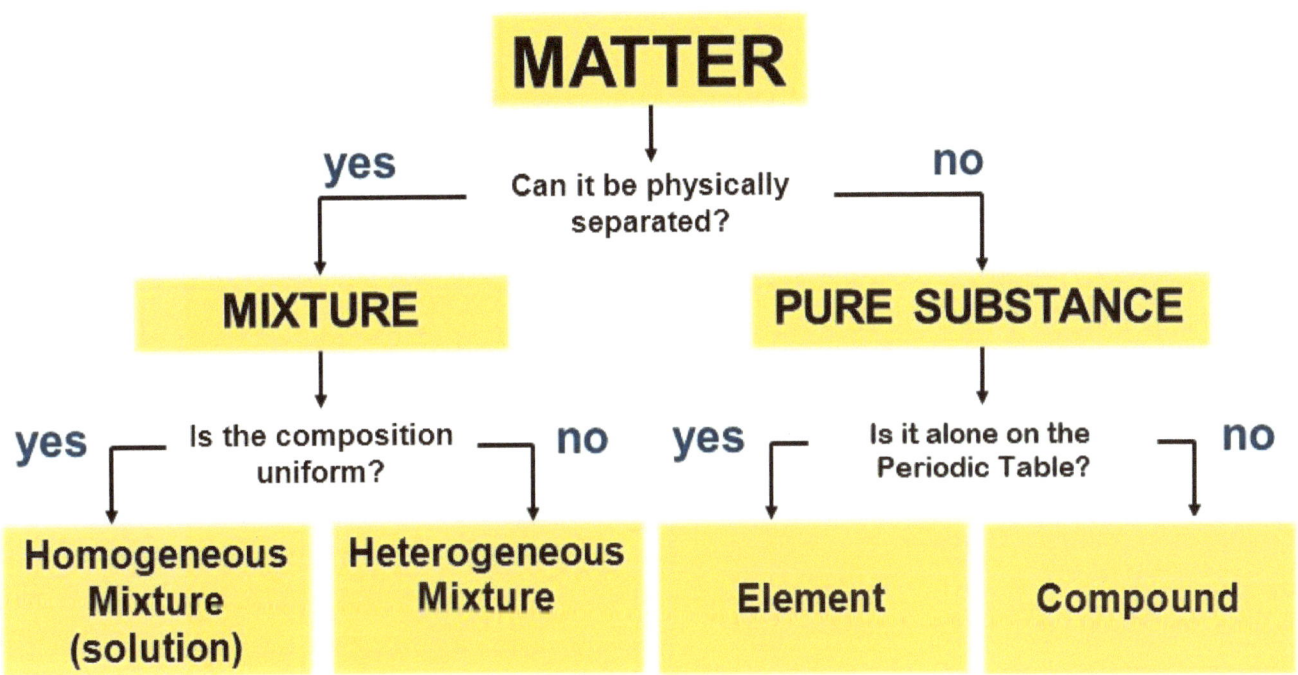

For the lecture, take an example bottle from the matter stations lab and talk your way through the chart with the class. This will be different (but similar) to the original lectures for each of the parts as we learned them and used pictures to demonstrate.

First, I hold up the bottle of water and ask, "Can it be separated physically?" Students will decide (give help if they need it), and you move to the **Pure Substance** box. "This is ALL water, so it is pure or all the same thing."

Then I ask, "Is WATER alone on the periodic table?" Going further, I ask, "What is the FORMULA for water?" They should know it is H_2O, which is Hydrogen and Oxygen, which are alone on the periodic table by themselves but not together as water when they are **chemically** combined. This is specifically my difference on this chart – students might be confused when you ask about being separated chemically (the typical question), but virtually all can look at the formula for the substance and know if it is an element (alone) or a compound (two or more elements). As a reinforcement, I pick a metal, such as the copper, and we do it again to get us an element.

Once you have the pure substance side down, choose a mixture bottle, such as the sulfur and iron bottle for heterogenous mixture. When I say this, I exaggerate the AND. We work our way down, and I ask about pouring out samples – uniform or not? If we did a good job on this yesterday, this part should be relatively easy, but you may have to hit it again with your extra bottles.

Finally, I do the solution (homogeneous mixture). I hold up a bottle of 'sugar water' or 'salt water' and ask, "What's in here?" They tell me, and I parrot back, "Salt AND water." Did you hear the AND? If there is an 'and' then it is always a mixture. All you have to decide is if you pour out samples, will they be uniform.

Once they have how to use the chart, they are ready to practice with the sorting stations. I tell them that each sample works the same way – look at what's in it and then go through the chart. You should remind them that we do NOT open the containers, please. Mine are not sealed, but if that is a problem with your students, you might consider hot glue to prevent any spills.

Activity - matter sorting stations. You will have at least one station per lab group. You can give the students a timer at each station and rotate or just have them move as a group from station to station as they finish it. You can also put the stations in a small basket and pass them around from group to group – I had to do that one year when we were 35+ in every class.

I like to turn them loose, so I have at least one extra station so they always have somewhere to move to until they have done them all completed. Once they figure out how to do it, they will finish quickly. Have fun with it as they explore and explain to each other how they use the chart and get what they get.

If you want to post the answers somewhere in the room, that is also a good idea. Once they finish, they go and grade their paper (I would hang out close to that location to prevent cheating) and turn them in or hold for a discussion if time permits. I grade on points, 1, 2 or 3 as always based on effort.

Matter Sorting Stations Setup:

You need samples for each of the four outcomes – at least 3 of each: homogeneous, heterogeneous, element and compound.

I have placed mine in small vials, but you can use any clear container (small jars, test tubes with stoppers, paint pots, etc.)

If you want to give the students a list for each station of what is in the bottles, you can provide that. If you want to increase the difficulty, have the students make their best guess as to what is in the bottle and then use the chart from there and leave the 'stuff' for the discussion later.

Number or letter each of the bottles so they are randomized on their answer sheet – this will help prevent them from deciding one and just filling in the rest without thinking it through.

I have one set of samples for the last group that are ORGANIC, which makes them more of a discussion set without a definite answer. Can organics be separated physically? Are they just a compound? Why did you put it where you did? Something for them to think about.

This is my set – you can use it for reference, but be creative, as almost anything will work!

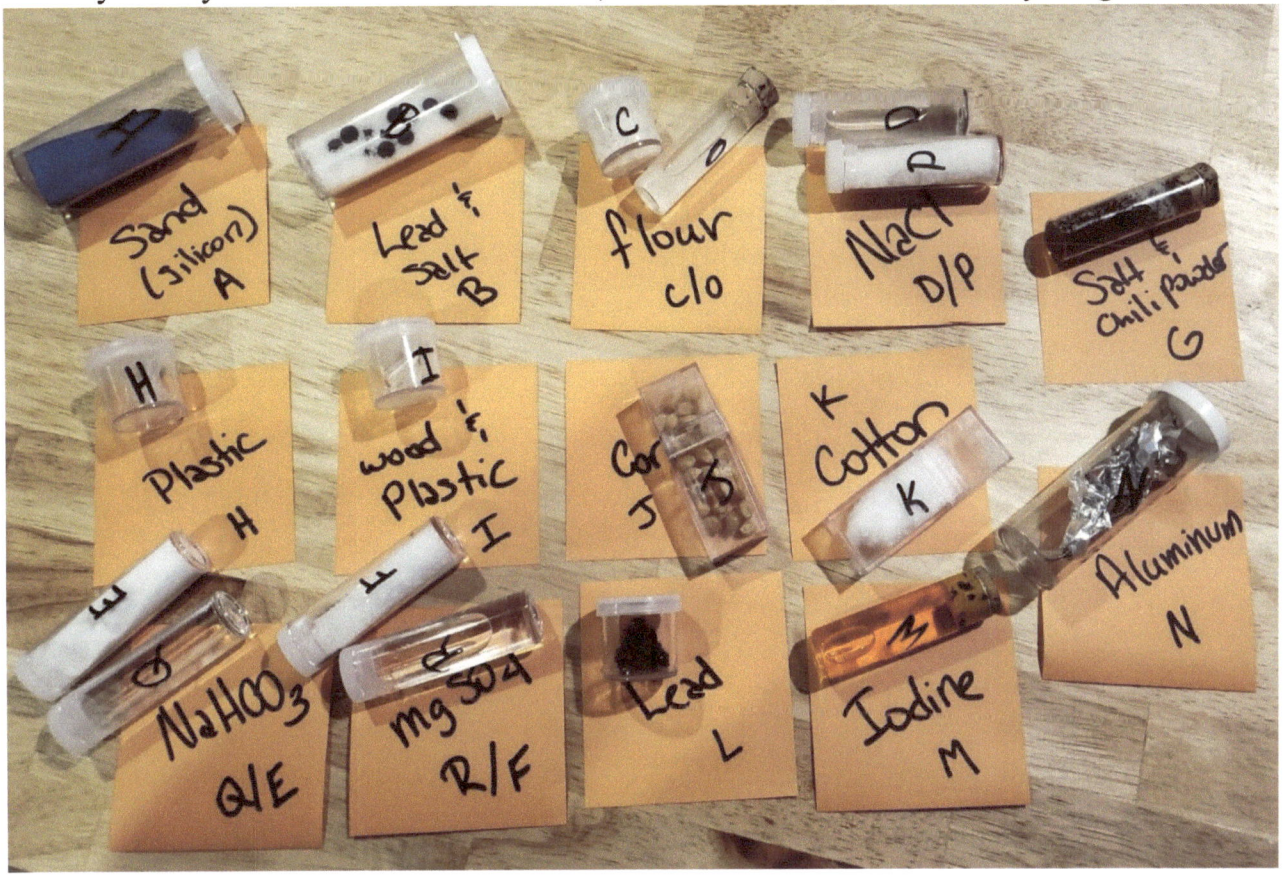

Topic: Properties of Matter Review
Day: 14
Unit: Properties of Matter

Learning Target:
I can review and practice concepts that will be covered on the Properties of Matter Exam.

Student Goals:
Differentiate between physical and chemical changes and properties.
Differentiate between intensive and extensive properties.
Differentiate between qualitative and quantitative properties
Classify measurements as precise and / or accurate.
Classify matter as pure substances or mixtures
Classify pure substances as elements or compounds
Classify mixtures as heterogenous or homogeneous

Agenda:
A – Matter Sorting Stations
A – Matter Change & Properties Cards
A – Matter Property Stations Lab
A – Review Sheet
A – Properties of Matter Vocabulary Quiz

Student Materials:
Journals for reviewing notes
Review Sheet copies for students to complete
Copies of the Properties of Matter Vocabulary Quiz

Props:

Actions and Rationale:
Warmup – students look over the review sheet while you take attendance so they can ask questions if they need to.

Activities – obviously these are some of the things we have done during the unit. Students can either make-up for when they were absent or revisit material to solidify their understanding with the review sheet to guide them. If you want to grade the review, I suggest taking it up the day of the exam so they can look over it until then. You can also use this day for the vocab quiz, which is short, or do it on another day.

Unit 2 Properties of Matter Test Review

1. Label the following as either Physical or Chemical Change?

Baking soda and vinegar react to form CO_2 _____   Tearing paper _____
Salt dissolving in water _____   Candle wax melting _____
Leaves changing color during the fall _____   Candle wick burning _____

2. Label the following as either Physical or Chemical Property?

Solubility _____   Flammability _____
Reactivity _____   Luster _____
Color _____   Ductility _____
Texture _____   Density _____

3. Accuracy and Precision: Draw a picture depicting accuracy and precision and define each one – be sure to have all 4 possible outcomes covered.

4. Qualitative measurements are _____

Give 3 examples for a glass of lemonade _____

5. Quantitative measurements are _____

Give 3 examples for a glass of lemonade _____

6.

Box A
mass=88 g

Boxes B & C
mass=44 g each

If the rock is 30 grams, what is the Density of the Rock ?

_____

Water only

Water after rock is placed in it

Density of Box A = _____   Density of Box B or C = _____

Would the boxes sink or float in water? _____ Why or Why not? _____

© Lavish Publishing, LLC

8. A ball has the mass of 2.5 grams and a volume of 5 mL. Calculate the density, then tell if it floats in water and how you know.

9. Label the following as either Intensive or Extensive Property.

Solubility _____   Flammability _____
Reactivity _____   Density _____
Color _____   Ductility _____
Mass _____   Volume _____

10. Label these as Pure Substance (element or Compound) or Mixture (Homogeneous/Heterogeneous)

11. Label these as pure substance (element or compound) or Mixture (homogeneous or heterogeneous)
    a. chocolate milk _____   d. Soured milk _____
    b. sugar water _____   e. carbon _____
    c. Olive garden Italian dressing _____   f. CO_2 _____

12. Draw and label the Matter Flow Chart

Mixtures can be categorized as _____ or _____

Pure Substances can be categorized as _____ or _____

© Lavish Publishing, LLC

Topic: Properties of Matter Unit Exam
Day: 15
Unit: Properties of Matter

Learning Target:
I can demonstrate my understanding of the Properties of Matter on a written exam.

Student Goals:
Differentiate between physical and chemical changes and properties.
Differentiate between intensive and extensive properties.
Differentiate between qualitative and quantitative properties
Classify measurements as precise and / or accurate.
Classify matter as pure substances or mixtures
Classify pure substances as elements or compounds
Classify mixtures as heterogenous or homogeneous

Agenda:
A – Red button day
A – Properties of Matter Exam

Student Materials:
Calculators
Scantrons or answer documents
Exam Copies

Actions and Rationale:
Red Button Day – classrooms run on routines, and this is one of my most hard fast. Basically, I have a red, yellow and green button up on my board next to the date, which is almost always on yellow. Yellow means that the students can be on their phones for APPROVED activities.

On red button day, ALL of their personal stuff goes in a designated location. I have a couple of folding tables at the front of my room that are set to about a foot tall, so they are short. Underneath, I have rugs because many kids don't like putting their bags on the floor underneath. Their bags, books, purses, etc. go on or under those tables and we do not move further until everyone has complied.

Their phones go either in their bag or on a charger, which I have two stations in my room – one that is a power strip where they use their charger, and a second set up with my personal chargers that I bought for the class to use – iPhone and android.

After we are seated and ready, I give them my testing rules –starting with the phones. We do the "pocket pat" so they can check to make sure their phones are put away and not accidentally back in their pocket, and I tell them what is going to happen if they don't. I like to have them stay seated and working until everyone is finished (I do allow them to slip out to the bathroom QUIETLY). I pick up all the exams at the end, and they are not allowed to get ANY of their stuff until I have all the exams in hand, and I dismiss them to get up and prepare to leave.

Phone penalty – if I see them with a phone – using it or not – until they have been dismissed, they get a 1 on the exam (a signal that they have broken testing rules). On the first offense, I let them take the exam again during tutorials the following week and contact their parents to let them know what happened and make the offer of a retake. After that, they get a 1 and it stands. This isn't a game, and they need to learn the consequences of breaking test security.

Yes, I am very strict on this. I am not naive, and I know that kids use their phones to cheat in an unlimited number of ways when given the chance. For this one small piece of my class, I want them to show me what they know on their own, and I don't want them sharing copies of my test in any way. If you are strict and diligent, the number of problems will be far less in the long run. Be clear and up front with your expectations and stick to your rules. The first few exams they may whine and complain, but it should die down. Anyone who is over the top or causes disruption of the test over it gets a phone call home so I can chat with their parents about why their student can't follow the rules like everyone else.

For the **answer documents**, we use scantrons, which makes grading super easy, but if you don't have access, you can always have them fill out their answers on a strip of paper, which will make them easier to grade. I also like to provide each student a copy of the exam, which I always print and shrink to fit on a single page. I like to do this so they can practice writing on the test and using test taking strategies as they work. I also keep them locked up before the test, and after they are used for the rest of the year in case we need to go back and look at someone's test, so I do ask them to put their names on them.

A note about **calculators**. I do not let the students use their phones as a calculator in my class. I make them use one from the set that I provide – all the same. They practice during class time before hand, and this is the only one they get on the test. This is a good habit and I suggest it to everyone whenever possible.

Property or Change Quiz

1. Is boiling water a property or a change?

2. Is the melting point of water a property or a change?

3. What is the difference between a property and a change?

© Lavish Publishing, LLC

Property or Change Quiz

1. Is boiling water a property or a change?

2. Is the melting point of water a property or a change?

3. What is the difference between a property and a change?

© Lavish Publishing, LLC

Property or Change Quiz

1. Is boiling water a property or a change?

2. Is the melting point of water a property or a change?

3. What is the difference between a property and a change?

© Lavish Publishing, LLC

Property or Change Quiz

1. Is boiling water a property or a change?

2. Is the melting point of water a property or a change?

3. What is the difference between a property and a change?

© Lavish Publishing, LLC

Chemical or Physical Quiz

1. Is melting point physical or chemical?

2. Is paper burning physical or chemical?

3. What is the difference between chemical and physical?

© Lavish Publishing, LLC

Chemical or Physical Quiz

1. Is melting point physical or chemical?

2. Is paper burning physical or chemical?

3. What is the difference between chemical and physical?

© Lavish Publishing, LLC

Chemical or Physical Quiz

1. Is melting point physical or chemical?

2. Is paper burning physical or chemical?

3. What is the difference between chemical and physical?

© Lavish Publishing, LLC

Chemical or Physical Quiz

1. Is melting point physical or chemical?

2. Is paper burning physical or chemical?

3. What is the difference between chemical and physical?

© Lavish Publishing, LLC

Intensive or Extensive Quiz

1. Is mass intensive or extensive?

2. Is color intensive or extensive?

3. What is the difference between intensive and extensive?

© Lavish Publishing, LLC

Intensive or Extensive Quiz

1. Is mass intensive or extensive?

2. Is color intensive or extensive?

3. What is the difference between intensive and extensive?

© Lavish Publishing, LLC

Intensive or Extensive Quiz

1. Is mass intensive or extensive?

2. Is color intensive or extensive?

3. What is the difference between intensive and extensive?

© Lavish Publishing, LLC

Intensive or Extensive Quiz

1. Is mass intensive or extensive?

2. Is color intensive or extensive?

3. What is the difference between intensive and extensive?

© Lavish Publishing, LLC

Quantitative or Qualitative Quiz

1. Is mass quantitative or qualitative?

2. Is color quantitative or qualitative?

3. What is the difference between quantitative and qualitative?

© Lavish Publishing, LLC

Quantitative or Qualitative Quiz

1. Is mass quantitative or qualitative?

2. Is color quantitative or qualitative?

3. What is the difference between quantitative and qualitative?

© Lavish Publishing, LLC

Quantitative or Qualitative Quiz

1. Is mass quantitative or qualitative?

2. Is color quantitative or qualitative?

3. What is the difference between quantitative and qualitative?

© Lavish Publishing, LLC

Quantitative or Qualitative Quiz

1. Is mass quantitative or qualitative?

2. Is color quantitative or qualitative?

3. What is the difference between quantitative and qualitative?

© Lavish Publishing, LLC

Precision and Accuracy Quiz

Label each target as Accurate, Precise, neither, or both.

A B C D

Precision and Accuracy Quiz

Label each target as Accurate, Precise, neither, or both.

A B C D

Precision and Accuracy Quiz

Label each target as Accurate, Precise, neither, or both.

A B C D

Density Quiz

1. A ball has a mass of 15 g and a volume of 3 mL. Calculate the density.

2. A ball has a density of .5 g/mL and a volume of 10 mL. Calculate the mass.

3. Which ball will float - the one in question 1 or question 2?

© Lavish Publishing, LLC

Density Quiz

1. A ball has a mass of 15 g and a volume of 3 mL. Calculate the density.

2. A ball has a density of .5 g/mL and a volume of 10 mL. Calculate the mass.

3. Which ball will float - the one in question 1 or question 2?

© Lavish Publishing, LLC

Density Quiz

1. A ball has a mass of 15 g and a volume of 3 mL. Calculate the density.

2. A ball has a density of .5 g/mL and a volume of 10 mL. Calculate the mass.

3. Which ball will float - the one in question 1 or question 2?

© Lavish Publishing, LLC

Density Quiz

1. A ball has a mass of 15 g and a volume of 3 mL. Calculate the density.

2. A ball has a density of .5 g/mL and a volume of 10 mL. Calculate the mass.

3. Which ball will float - the one in question 1 or question 2?

© Lavish Publishing, LLC

Element or Compound Quiz

1. Is this a picture of an element or compound?

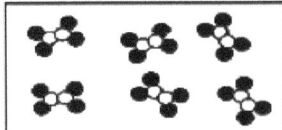

2. Is this a picture of an element or compound?

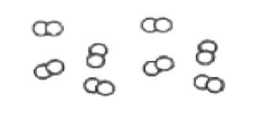

3. What is the difference between an element and a compound?

© Lavish Publishing, LLC

Element or Compound Quiz

1. Is this a picture of an element or compound?

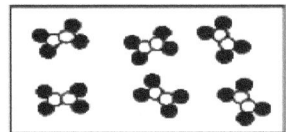

2. Is this a picture of an element or compound?

3. What is the difference between an element and a compound?

© Lavish Publishing, LLC

Element or Compound Quiz

1. Is this a picture of an element or compound?

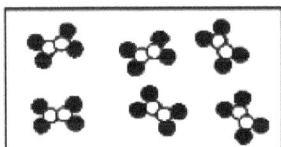

2. Is this a picture of an element or compound?

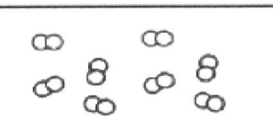

3. What is the difference between an element and a compound?

© Lavish Publishing, LLC

Element or Compound Quiz

1. Is this a picture of an element or compound?

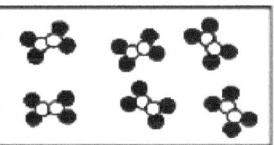

2. Is this a picture of an element or compound?

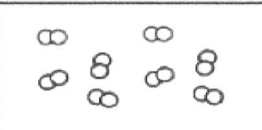

3. What is the difference between an element and a compound?

© Lavish Publishing, LLC

Pure Substance & Mixture Quiz

1. Is this a picture of
a mixture or pure substance?

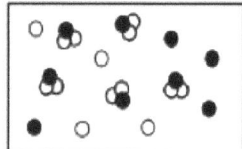

2. Is this a picture of
a mixture or pure substance?

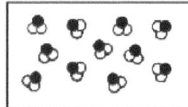

3. What is the difference between
a mixture and a pure substance?

© Lavish Publishing, LLC

Pure Substance & Mixture Quiz

1. Is this a picture of
a mixture or pure substance?

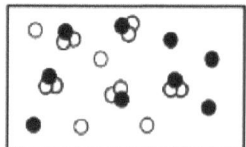

2. Is this a picture of
a mixture or pure substance?

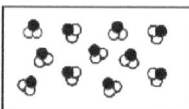

3. What is the difference between
a mixture and a pure substance?

© Lavish Publishing, LLC

Pure Substance & Mixture Quiz

1. Is this a picture of
a mixture or pure substance?

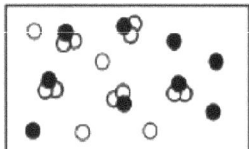

2. Is this a picture of
a mixture or pure substance?

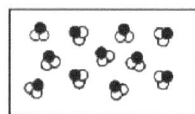

3. What is the difference between
a mixture and a pure substance?

© Lavish Publishing, LLC

Pure Substance & Mixture Quiz

1. Is this a picture of
a mixture or pure substance?

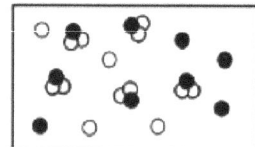

2. Is this a picture of
a mixture or pure substance?

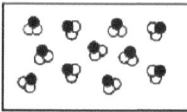

3. What is the difference between
a mixture and a pure substance?

© Lavish Publishing, LLC

Heterogeneous & Homogeneous Quiz

1. Is this a mixture homogeneous or heterogeneous?

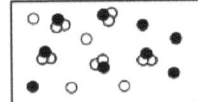

2. Is this a mixture homogeneous or heterogeneous?

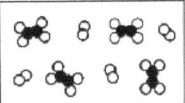

3. What is the difference between homogeneous and heterogeneous?

© Lavish Publishing, LLC

Heterogeneous & Homogeneous Quiz

1. Is this a mixture homogeneous or heterogeneous?

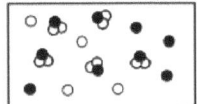

2. Is this a mixture homogeneous or heterogeneous?

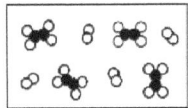

3. What is the difference between homogeneous and heterogeneous?

© Lavish Publishing, LLC

Heterogeneous & Homogeneous Quiz

1. Is this a mixture homogeneous or heterogeneous?

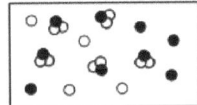

2. Is this a mixture homogeneous or heterogeneous?

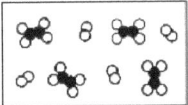

3. What is the difference between homogeneous and heterogeneous?

© Lavish Publishing, LLC

Heterogeneous & Homogeneous Quiz

1. Is this a mixture homogeneous or heterogeneous?

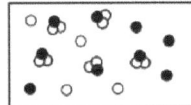

2. Is this a mixture homogeneous or heterogeneous?

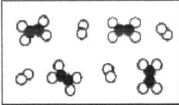

3. What is the difference between homogeneous and heterogeneous?

© Lavish Publishing, LLC

Matter Flowchart Quiz

Fill in the missing boxes

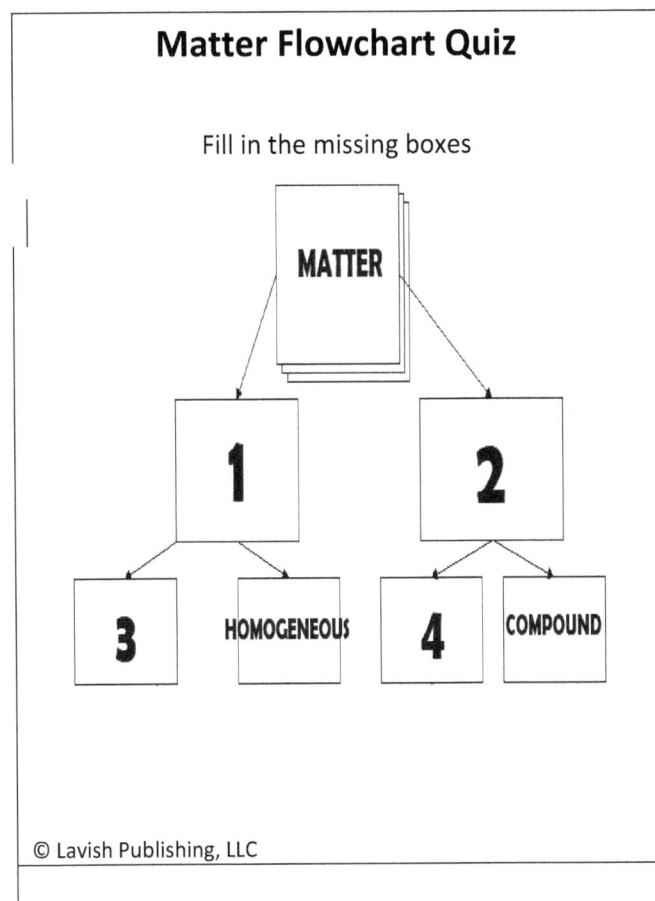

© Lavish Publishing, LLC

Matter Flowchart Quiz

Fill in the missing boxes

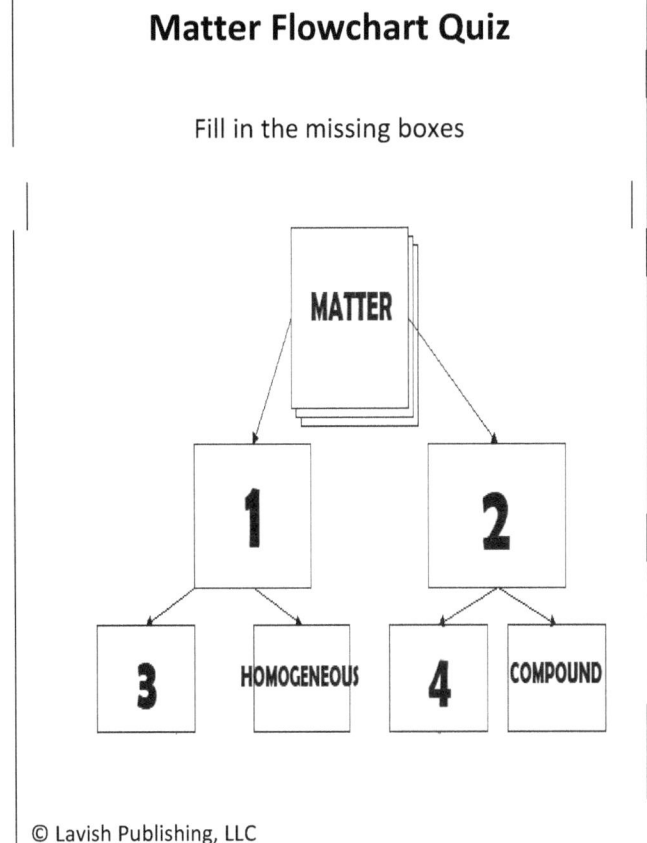

© Lavish Publishing, LLC

Matter Flowchart Quiz

Fill in the missing boxes

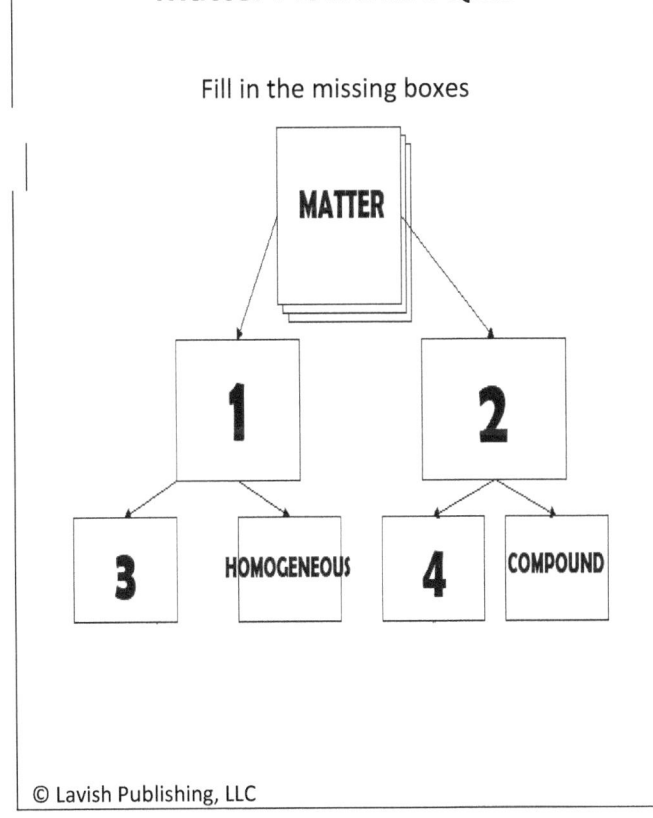

© Lavish Publishing, LLC

Matter Flowchart Quiz

Fill in the missing boxes

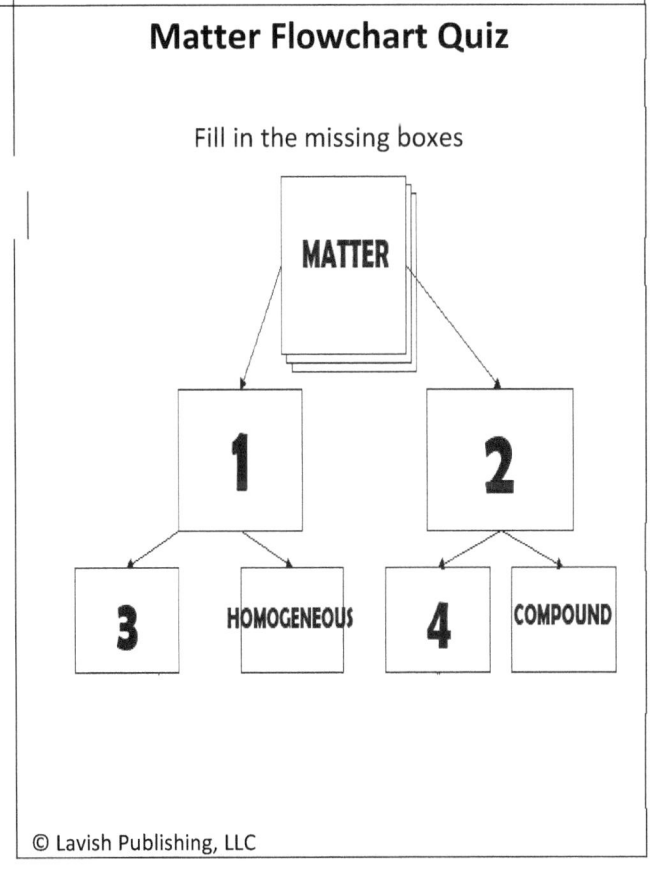

© Lavish Publishing, LLC

From A Properties of Matter Vocab Quiz
© Lavish Publishing, LLC

1		___ is matter in its simplest form.
2		___ is the characteristic physical property referring to the ability of a given substance, the solute, to dissolve in a solvent. It is measured in terms of the maximum amount of solute dissolved in a solvent at equilibrium.
3		___ is the chemical property of a compound that is poisonous or harmful if ingested (consumed).
4		___ is calculated to determine how close or far our results are from where they should be based on the accepted value.
5		___ properties or values that can NOT be given number form - such as color.
	word choices	(A) element (B) percent error (C) qualitative (D) solubility (E) toxicity

Form B Properties of Matter Vocab Quiz
© Lavish Publishing, LLC

1		A ____ mixture has unvaried amounts when sampled; separate physically.
2		_____ is an intensive property of a substance that is measured in g per ml. It is a factor that determines whether an object will sink or float when placed in a liquid.
3		_____ is the ability of metal to be shaped or flattened.
4		_____ is the amount of space matter occupies.
5		A ___ is a solid formed from combining two liquids - indicates there was a chemical reaction.
	word choices	(A) density (B) homogeneous (C) malleability (D) precipitate (E) volume

Form C Properties of Matter Vocab Quiz
© Lavish Publishing, LLC

1		A ____ mixture has varied amounts when sampled; separate physically.
2		A ____ property can be observed without changing the composition of the sample.
3		A ____ is 2 or more different elements chemically combined.
4		A change from one state (solid or liquid or gas) to another without a change in chemical composition is called a ____.
5		An ____ property stays the same regardless of sample size - like density.
	word choices	(A) compound (B) heterogeneous (C) intensive (D) phase change (E) physical

Form D Properties of Matter Vocab Quiz
© Lavish Publishing, LLC

1		The willingness of an element to form bonds is called ____.
2		A ____ mixture has unvaried amounts when sampled; separate physically.
3		An ____ property changes with the size of our sample - example is volume.
4		The book value or one given to you by you teacher is called the ____; it is what we should be getting in our measurements.
5		Examples of ____ would be a cup of sugar, a flask of HCL, or a bottle of O_2, which means only a single substance is present.
	word choices	(A) Accepted value (B) extensive (C) homogeneous (D) pure substance (E) reactivity

Properties and Changes in Matter Exam

1. When milk sours or rots, that would be…
   A. Chemical Property
   B. Chemical Change
   C. Physical Property
   D. Physical Change

2. When you grind or crush pepper, that would be…
   A. Chemical Property
   B. Chemical Change
   C. Physical Property
   D. Physical Change

3. The flammability of paper (ability to burn) would be…
   A. Chemical Property
   B. Chemical Change
   C. Physical Property
   D. Physical Change

4. The malleability of metal (ability to bend) would be…
   A. Chemical Property
   B. Chemical Change
   C. Physical Property
   D. Physical Change

5. Which would be an intensive property?
   A. dimensions
   B. color
   C. mass
   D. volume

6. Which would be an extensive property?
   A. density
   B. color
   C. shape
   D. volume

7. Which properties would be quantitative?
   A. density
   B. color
   C. shape
   D. flavor

8. Which properties would be qualitative?
   A. density
   B. mass
   C. color
   D. volume

9. Which of the following is a correct statement regarding mixtures?
   A. Mixtures have a fixed composition.
   B. All substances in a mixture are visible at all times.
   C. Mixtures are composed of a single substance.
   D. Mixtures are composed of more than one substance.

© Lavish Publishing, LLC

10. Which of the following is a correct statement regarding pure substances?
    A. Pure substances are composed of a single substance.
    B. Pure substances are composed of more than one substance.
    C. Pure substances can never be separated in any way.
    D. Pure substances can be mixtures.

Observe the chart for separating matter.

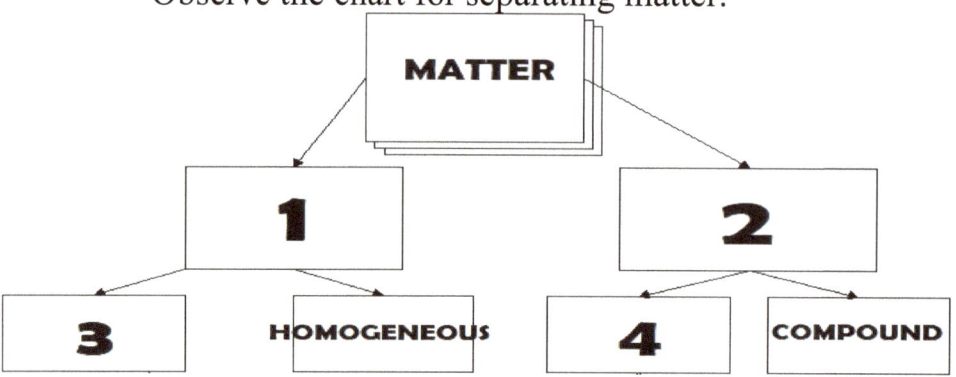

11. Which is the box for Pure Substance?
    A. 1   C. 3
    B. 2   D. 4

12. Which is the box for element?
    A. 1   C. 3
    B. 2   D. 4

13. Which target demonstrates accuracy and precision?

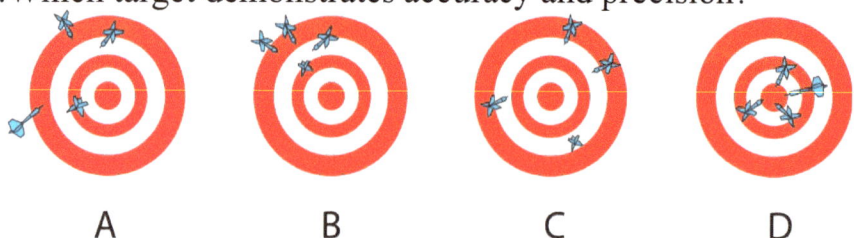

A   B   C   D

14. Which target demonstrates precision but not accuracy?

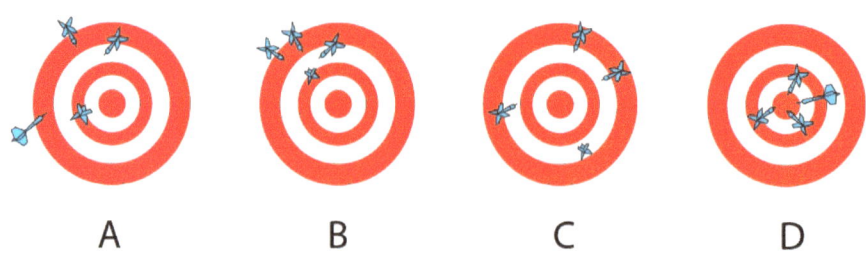

A   B   C   D

© Lavish Publishing, LLC

15. Calculate the density of a block that is 2 cm x 2 cm x 2 cm with a mass of 48g.
    A. 3 g/cm3
    B. 6 g/cm3
    C. 12 g/cm3
    D. 24 g/cm3

16. A cube of copper is cut in half. The density of the cube was 12 g/cm³. What will the density of Section 1 be?

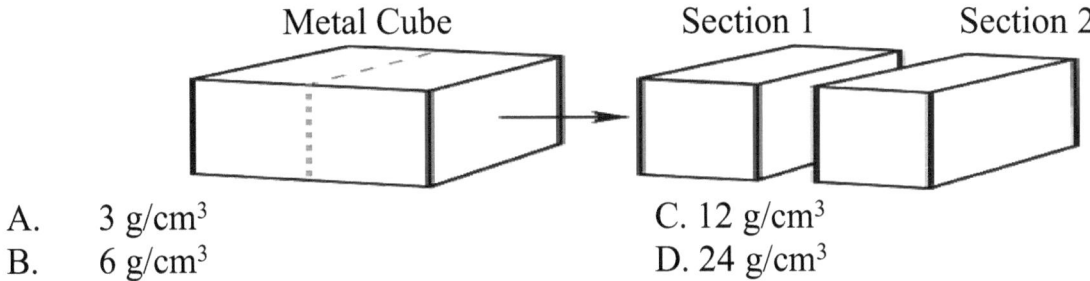

    A. 3 g/cm³
    B. 6 g/cm³
    C. 12 g/cm³
    D. 24 g/cm³

17. You have a collection of small balls, which you drop into a glass of water. One of the balls floats while the others sink. What do you know about the ball that floats?
    A. it has a density greater than the other balls
    B. it has a density greater than 1
    C. it has a density less than 1
    D. nothing can be determined from this experiment

18. You have a rubber ball with a mass of 3 grams and a density of 0.5 g/mL. What is the volume of the ball?
    A. 3 mL
    B. 6 mL
    C. 9 mL
    D. 24 mL

19. Which is an element?

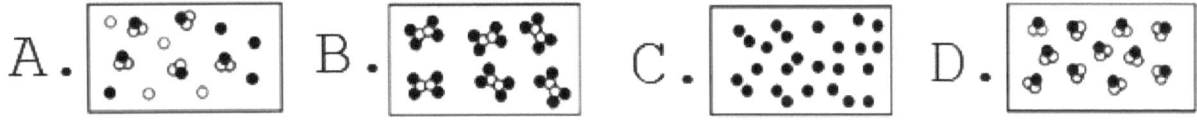

20. Which is a heterogeneous mixture?

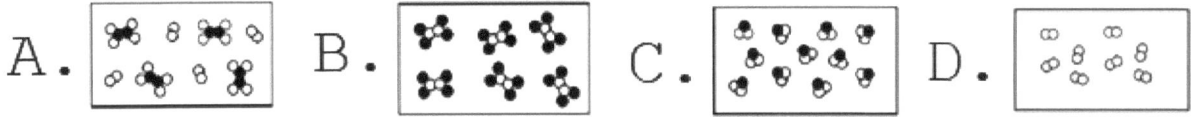

© Lavish Publishing, LLC

Quizzes Answers

Property or Change Quiz

1. Is boiling water a property or a change?
change - liquid to gas

2. Is the melting point of water a property or a change?
property - it's something water owns

3. What is the difference between a property and a change?
changes have a before and after
properties belong to the item - an ability

© Lavish Publishing, LLC

Chemical or Physical Quiz

1. Is melting point physical or chemical?
physical - it's on the states chart

2. Is paper burning physical or chemical?
chemical - see chemical change indicators

3. What is the difference between chemical and physical?
physical means the substance is the same,
chemicial mean a new substance forms

© Lavish Publishing, LLC

Intensive or Extensive Quiz

1. Is mass intensive or extensive?
extensive - changes with sample size

2. Is color intensive or extensive?
intensive - not affect by sample size

3. What is the difference between intensive and extensive?
intensive is permanent, like identity
extensive is changed by sample size

© Lavish Publishing, LLC

Quantitative or Qualitative Quiz

1. Is mass quantitative or qualitative?
quantitative - expressed as a number

2. Is color quantitative or qualitative?
qualitiative - not expressed as a number

3. What is the difference between quantitative and qualitative?
qualitative values are like 'qualities'
quantitative can be quantified as a number

© Lavish Publishing, LLC

Density Quiz

1. A ball has a mass of 15 g and a volume of 3 mL. Calculate the density.
5 g/mL

2. A ball has a density of .5 g/mL and a volume of 10 mL. Calculate the mass.
5 g

3. Which ball will float - the one in question 1 or question 2?
2 - it's density is less than 1

© Lavish Publishing, LLC

Element or Compound Quiz

1. Is this a picture of an element or **compound**?

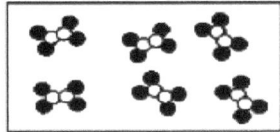

2. Is this a picture of an **element** or compound?

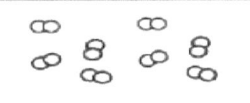

3. What is the difference between an element and a compound?
elements are alone on the periodic table
compounds are made of elements -
2 more _different_ elements

© Lavish Publishing, LLC

Pure Substance & Mixture Quiz

1. Is this a picture of a **mixture** or pure substance?

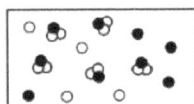

2. Is this a picture of a mixture or **pure substance**?

3. What is the difference between a mixture and a pure substance?
mixtures are physically combined
pure substances are all the same thing,
(could be elements or compounds)

Heterogeneous & Homogeneous Quiz

1. Is this a mixture **homogeneous** or heterogeneous?

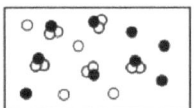

2. Is this a mixture homogeneous or **heterogeneous**?

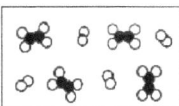

3. What is the difference between homogeneous and heterogeneous?
homogeneous is a solution - something
dissolved in water with uniform samples
heterogeneous has varied samples

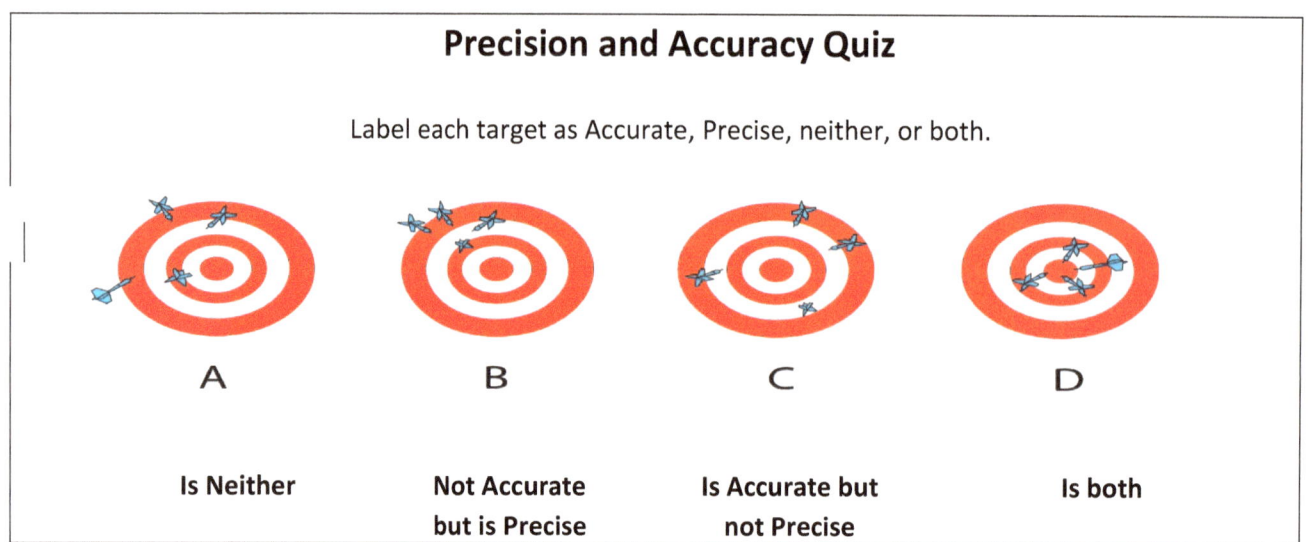

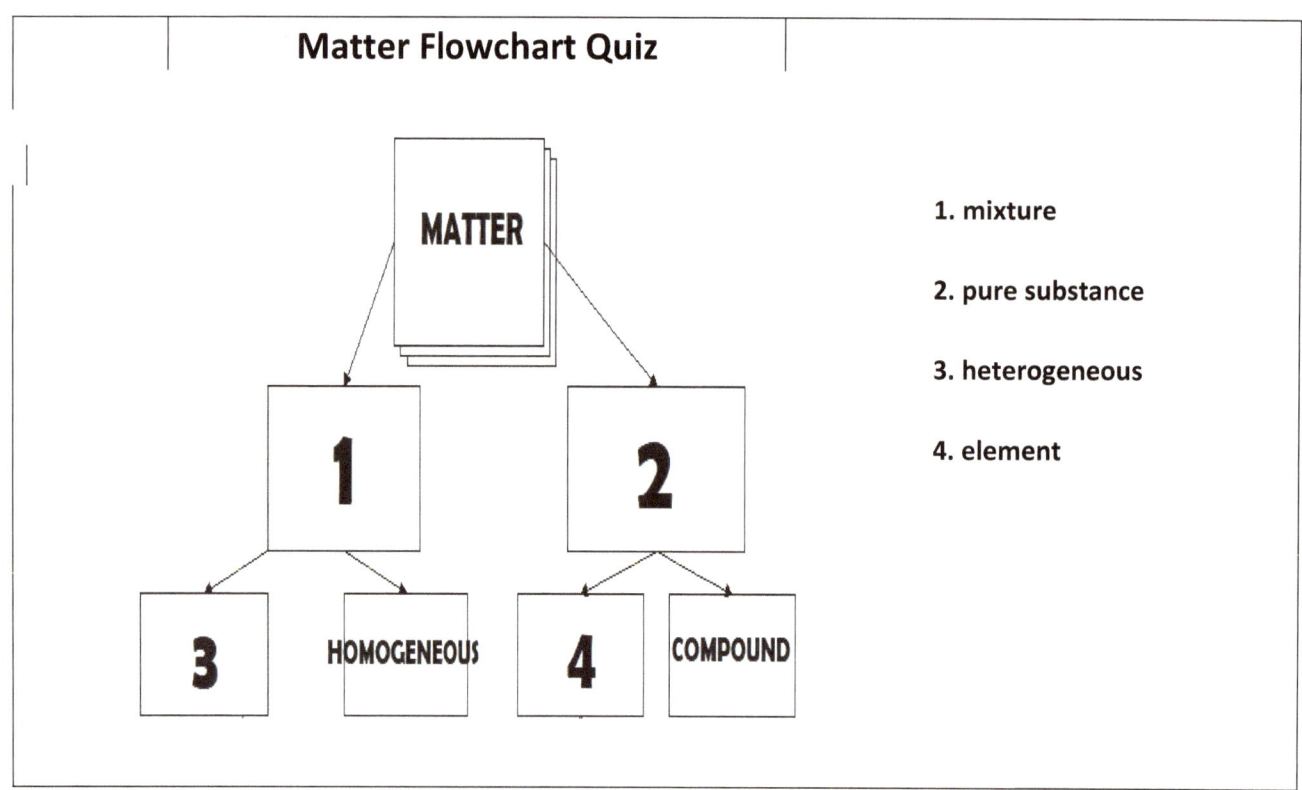

1. mixture

2. pure substance

3. heterogeneous

4. element

Properties of Matter Vocabulary Quiz Answers

Form A		Form B		Form C		Form D	
1	A Element	1	B homogenous	1	B heterogeneous	1	E reactivity
2	D Solubility	2	A density	2	E physical	2	C homogenous
3	E Toxicity	3	C malleability	3	A compound	3	B extensive
4B	B Percent error	4	E volume	4	D phase change	4	A accepted value
5	C qualitative	5	D precipitate	5	C intensive	5	D pure substance

Properties of Matter Exam Answers

1. B   6. D   11. B   16. C
2. D   7. A   12. D   17. C
3. A   8. C   13. D   18. B
4. C   9. D   14. B   19. C
5. B   10. A   15. B   20. A

About the Author

Born and raised in West Texas, Sammie Jacobs aspired to be a teacher at an early age but did not achieve her dream until the age of 38 when she earned her composite science certification in 2008.

Answering the call of the classroom, she went to work at a local high school teaching Chemistry. Through all the years, she loved the subject and her students. Combining her dedication to both, Sam continuously searched for ways to improve her instruction and meet her students varied needs, leading her to create and construct many of her materials.

Mentoring new teachers as they came into the district and her department, Sam could see the importance of helping those new to her beloved field. First, she began to build the files that would become the foundation of this series. Later, she realized that putting those lesson plans and tools into the hands of others could mean a great deal to her colleagues and countless students, eventually even those across the country or around the world.

Always striving for more and looking for the best in herself and those around her, Sammie Jacobs is releasing this complete version of her Chemistry lessons – 17 units in all, which will be available for purchase at a nominal fee in paperback format. These files contain everything needed to plan and execute a solid foundational year of Chemistry for any teacher, be it veteran or novice.

But Sam also knows there is more that can be done, so she is also founding a teacher community group on Facebook. Open to everyone who works as a science educator, this is Sam's legacy, her dream coming true. If you are a veteran with advice to share or a novice looking for a helpful hand, come and join and let us grow together. In the end, Sam hopes many will find these tools useful and many more will be inspired to reach for their dreams, whatever they might be…

Units in the Teaching Chemistry in a Diversified Classroom Series

Lab Safety & Equipment
Properties of Matter
Periodic Table
Atomic Structure
Electrons in Atoms
Ionic Bonding
Covalent Bonding
Names & Formulas
Equations & Reactions
Calculations
Stoichiometry
States of Matter
Gases
Solutions
Acids & Bases
Thermochemistry
Nuclear Chemistry

Follow the Teaching Chemistry Series on Facebook –
https://www.facebook.com/SammieJacobsChemistry/?

Join the Chemistry Professional Learning Community on Facebook –
https://www.facebook.com/groups/TeachingChemistryPLC/

www.ingramcontent.com/pod-product-compliance
Lightning Source LLC
Chambersburg PA
CBHW042034150426
43201CB00002B/25